Climate Change and its Causes, Effects and Prediction Series

DESIGNING GREENHOUSE GAS REDUCTION AND REGULATORY SYSTEMS

CLIMATE CHANGE AND ITS CAUSES, EFFECTS AND PREDICTION SERIES

Global Climate Change
Horace M. Karling (Editor)
2001 ISBN: 1-56072-999-6

Global Climate Change Revisited
Horace M. Karling (Editor)
2007 ISBN: 1-59454-039-X

Climate Change Research Progress
Lawrence N. Peretz (Editor)
2008 ISBN: 1-60021-998-5

Climate Change: financial Risks
United States Government Accountability Office
2008 ISBN: 978-1-60456-488-4

Post-Kyoto: Designing the Next International Climate Change Protocol
Matthew Clarke
2008 ISBN: 978-1-60456-840-0

The Effects of Climate Change on Agriculture, Land Resources, water Resources, and Biodiversity in the United States
Peter Backlund, Anthony Janetos and David Schimel
2009 ISBN: 978-1-60456-989-6

Economics of Policy Options to address Climate Change
Gregory N. Bartos
2009 ISBN: 978-1-60692-116-6

Climate Variability, Modeling Tools and Agricultural Decision-Making
Angel Utset (Editor)
2009 ISBN: 978-1-60692-703-8

Emissions Trading: Lessons Learned from the European Union and Kyoto Protocol Climate Change Programs
Ervin Nagy and Gisella Varga (Editors)
2009 ISBN: 978-1-60741-194-9

Global Climate Change: International Perspectives and Responses
Elias D'Angelo (Editor)
2009 ISBN: 978-60741-233-5

Disputing Global Warming
Anton Horvath and Boris Molnar (Editors)
2009 ISBN: 978-1-60741-235-9

Climate Change and its Causes, Effects and Prediction Series

DESIGNING GREENHOUSE GAS REDUCTION AND REGULATORY SYSTEMS

SONJA ENDEN
EDITOR

Nova Science Publishers, Inc.
New York

Copyright © 2009 by Nova Science Publishers, Inc.

All rights reserved. No part of this book may be reproduced, stored in a retrieval system or transmitted in any form or by any means: electronic, electrostatic, magnetic, tape, mechanical photocopying, recording or otherwise without the written permission of the Publisher.

For permission to use material from this book please contact us:
Telephone 631-231-7269; Fax 631-231-8175
Web Site: http://www.novapublishers.com

NOTICE TO THE READER

The Publisher has taken reasonable care in the preparation of this book, but makes no expressed or implied warranty of any kind and assumes no responsibility for any errors or omissions. No liability is assumed for incidental or consequential damages in connection with or arising out of information contained in this book. The Publisher shall not be liable for any special, consequential, or exemplary damages resulting, in whole or in part, from the readers' use of, or reliance upon, this material.

Independent verification should be sought for any data, advice or recommendations contained in this book. In addition, no responsibility is assumed by the publisher for any injury and/or damage to persons or property arising from any methods, products, instructions, ideas or otherwise contained in this publication.

This publication is designed to provide accurate and authoritative information with regard to the subject matter covered herein. It is sold with the clear understanding that the Publisher is not engaged in rendering legal or any other professional services. If legal or any other expert assistance is required, the services of a competent person should be sought. FROM A DECLARATION OF PARTICIPANTS JOINTLY ADOPTED BY A COMMITTEE OF THE AMERICAN BAR ASSOCIATION AND A COMMITTEE OF PUBLISHERS.

LIBRARY OF CONGRESS CATALOGING-IN-PUBLICATION DATA
Available Upon Request

ISBN: 978-1-60741-195-6

Published by Nova Science Publishers, Inc. ✦ New York

CONTENTS

Preface		**vii**
Chapter 1	Climate Change: Design Approaches for Greenhouse Gas Reduction Program *Larry Parker*	1
Chapter 2	Comments on Design Elements of a Mandatory Market-Based Greenhouse Gas Regulatory System	33
Chapter 3	Design Elements of a Mandatory Market-Based Greenhouse Gas Regulatory System *Pete V. Domenici and Jeff Bingaman*	59
Index		73

PREFACE

This book explores some of the key questions and design elements of a national greenhouse gas program in order to facilitate discussion and the development of consensus around a specific bill. There are many ways to structure a regulatory program and different approaches that might include a carbon tax, technology incentives and voluntary programs, but consideration is limited in this report to "mandatory market-based systems" contemplated by the Sense of the Senate Resolution.

Chapter 1 - Two events provide impetus for revisiting the cost issue with respect to designing a greenhouse gas reduction program. The first is the election of a new President publicly committed to substantial reductions in greenhouse gases over the next several decades. The second was passage of the 2005 Sense of the Senate climate change resolution calling on the Congress to enact a mandatory, market- based program to slow, stop, and reverse the growth of greenhouse gases, and which states that the program should be enacted at a rate and in a manner that "will not significantly harm the United States economy" and "will encourage comparable action" by other nations. Facets of the cost issue that have raised concern include absolute costs to the economy, distribution of costs across industries, competitive impact domestically and internationally, incentives for new technology, and uncertainty about possible costs.

Chapter 2 - In light of concerns that rising concentrations of greenhouse gases in the atmosphere may be affecting the Earth's climate, several Members of Congress and public interest groups have proposed plans to require cuts in the United States' emissions of those gases. Implementing a "cap-and-trade" program is an example of one such proposal. Under such a program, policymakers would establish an overall cap on emissions but allow regulated firms to trade rights to those emissions, called allowances. That trading would provide an incentive for firms that could reduce their emissions most cheaply to sell some of their allowances to firms that faced higher costs to reduce their emissions. Such an approach would help reduce the costs of achieving the emissions cap.

Chapter 3 - The purpose of this document is to lay out some of the key questions and design elements of a national greenhouse gas program in order to facilitate discussion and the development of consensus around a specific bill. We recognize that there are many ways to structure such a regulatory program and that there are entirely different approaches that might include a carbon tax, technology incentives and voluntary programs, but we have limited our consideration here to "mandatory market-based systems" contemplated by the Sense of the Senate Resolution.

Chapter 1

CLIMATE CHANGE: DESIGN APPROACHES FOR GREENHOUSE GAS REDUCTION PROGRAM*

Larry Parker

ABSTRACT

Two events provide impetus for revisiting the cost issue with respect to designing a greenhouse gas reduction program. The first is the election of a new President publicly committed to substantial reductions in greenhouse gases over the next several decades. The second was passage of the 2005 Sense of the Senate climate change resolution calling on the Congress to enact a mandatory, market- based program to slow, stop, and reverse the growth of greenhouse gases, and which states that the program should be enacted at a rate and in a manner that "will not significantly harm the United States economy" and "will encourage comparable action" by other nations. Facets of the cost issue that have raised concern include absolute costs to the economy, distribution of costs across industries, competitive impact domestically and internationally, incentives for new technology, and uncertainty about possible costs.

Market-based mechanisms address the cost issue by introducing flexibility into the implementation process. The cornerstone of that flexibility is permitting sources to decide for themselves their appropriate

---

* This is an edited, excerpted and augmented edition of a Congressional Research Service publication Report RL33799, dated November 24, 2008.

implementation strategy within the parameters of market signals and other incentives. That signal can be as simple as a carbon tax or comprehensive credit auction that tells the emitter the value of any reduction in greenhouse gases, to a credit marketplace that is constrained by a ceiling price (safety valve) and includes incentives for new technology. As illustrated here, the combinations of market mechanisms are numerous, allowing decision makers to tailor the program to address specific concerns.

In general, market-based mechanisms to reduce greenhouse gas emissions, the most important being carbon dioxide (CO_2), focus on specifying either the acceptable emissions level (quantity) or the compliance costs (price), and allowing the marketplace to determine the economically efficient solution for the other variable. For example, a tradeable permit program sets the amount of emissions allowable under the program (i.e., the number of permits available limits or caps allowable emissions), while allowing the marketplace to determine what each permit will be worth. Likewise, a carbon tax sets the maximum unit cost (per ton of CO_2 equivalent) that one should pay for reducing emissions, while the marketplace determines how much actually gets reduced.

In one sense, preference for a carbon tax or a tradeable permit system depends on how one views the uncertainty of costs involved and benefits to be received. The options discussed here represent a continuum between alternatives focused on the price side of the equation (e.g., carbon taxes) through hybrid schemes (e.g., safety valves) to alternatives focused on the quantity side (e.g., banking and borrowing). They are tools to assist in the assessment of potential greenhouse gas reduction approaches, leaving any policy decision on balancing the price-quantity issue to the ultimate decision makers.

CLIMATE CHANGE: DESIGN APPROACHES FOR A GREENHOUSE GAS REDUCTION PROGRAM

Climate change has been a continuing policy issue since the United States ratified the 1992 United Nations Framework Convention on Climate Change (UNFCCC). An integral part, and sometimes driving part, of the ensuing debate has been the issue of cost — in several manifestations. [1] For the George W. Bush Administration, the Kyoto Protocol was "fatally flawed in fundamental ways," including requiring compliance with mandates that "would have a negative economic impact with layoffs of workers and price increases for consumers." [2] This concern about cost can also be seen in the Senate's most recent resolution on climate change in 2005 (S.Amdt. 866). Echoing the language of its 1997 resolution (S.Res. 98) on the same subject, the 2005 Sense of the Senate resolution on climate change declared that a mandatory, market-based program to slow, stop, and reverse the growth of greenhouse gases [3] should be enacted at a rate and in a manner that "will not significantly harm the United States economy" and "will encourage comparable action" by other nations. [4] Facets of the cost issue that have raised concern include absolute costs to the economy, distribution of costs across industries, competitive impact domestically and internationally, incentives for new technology, and uncertainty about possible costs.

Because a stalemate has persisted on strategies to control greenhouse gas (GHG) emissions, particularly because of cost uncertainties, attention has increasingly focused on options to address these concerns and to move the debate forward. These options range from incremental mechanisms within a tradeable permit program, such as banking and borrowing of credits, which minimally affect overall emissions reduction targets, to more fundamental proposals, such as a carbon tax, which would take climate change policy in a new and somewhat uncharted direction.

This paper explores these options to address the cost issue in four parts. First, the basic economic tradeoff between controlling the quantity of GHG emissions and the program's compliance costs is introduced and explained. Second, the five dimensions of the cost issue that have arisen so far in the climate change debate are identified and discussed. Third, a representative sample of proposed approaches to address cost concerns is compared and analyzed according to the five cost dimensions identified previously. Finally, the proposed options are summarized and opportunities to combine or merge different approaches are analyzed. The paper does not provide a detailed discussion of allocation and implementation issues that creating a market-based mechanism (particularly a cap-and-trade program) would entail.

INTRODUCTION: THE PRICE VERSUS QUANTITY DEBATE

In general, market-based mechanisms to reduce GHG emissions, the most important being carbon dioxide (CO_2), focus on specifying either the acceptable emissions level (quantity) or the compliance costs (price) and allowing the marketplace to determine the economically efficient solution for the other variable. For example, a tradeable permit program sets the amount of emissions allowable under the program (i.e., the number of permits available limits or caps allowable emissions), while permitting the marketplace to determine what each permit will be worth. Likewise, a carbon tax sets the maximum unit cost (per ton of CO_2 equivalent) that one should pay for reducing emissions, while the marketplace determines how much actually gets reduced. In one sense, preference for a carbon tax or a tradeable permit system depends on how one views the uncertainty of costs involved and benefits to be received.

For those confident that achieving a specific level of CO_2 reduction will yield significant benefits — enough so that even the potentially very high end of the marginal cost curve does not bother them — a tradeable permit program may be most appropriate. CO_2 emissions would be reduced to a specific level, and in the case of a tradeable permit program, the cost involved would be handled efficiently, though not controlled at a specific cost level. This efficiency occurs because through the trading of permits, emissions reduction efforts focus on sources at which controls can be achieved at least cost.

However, if one feels uncertain of the environmental benefits of a specific level of reduction and anxious about the downside risk of substantial control costs to the economy, then a carbon tax may be most appropriate. In this approach, the level of the tax effectively caps the marginal cost of control that affected activities would pay under the reduction scheme, but the precise level of CO_2 reduction achieved is less certain. Emitters of CO_2 would spend money controlling CO_2 emissions up to the level of the tax. However, because the marginal cost of control among millions of emitters is not well known, the overall emissions reductions for a given tax level on CO_2 emissions is subject to some uncertainty.

Hence, a major policy question is whether one is more concerned about the possible economic cost of the program and therefore willing to accept some uncertainty about the amount of reduction received (i.e., carbon taxes); or one is more concerned about achieving a specific emissions reduction level with costs handled efficiently, but not capped (i.e., tradeable permits).

A model for a tradeable permit approach is the sulfur dioxide (SO_2) allowance program contained in Title IV of the 1990 Clean Air Act Amendments (42 U.S.C. 7651). Also called the Acid Rain Program, the tradeable permit system is based on two premises. First, a set amount of SO_2 emitted by human activities can be assimilated by the ecological system without undue harm. Thus the goal of the program is to put a ceiling, or cap, on the total emissions of SO_2 rather than limit ambient concentrations. Second, a market in pollution licenses between polluters is the most cost-effective means of achieving a given reduction. This market in pollution licenses (or allowances, each of which is equal to 1 ton of SO_2) is designed so that owners of allowances can trade those allowances with other emitters who need them or retain (bank) them for future use or sale. Initially, most allowances were allocated free by the federal government to utilities according to statutory formulas related to a given facility's historic fuel use and emissions; other allowances have been reserved by the government for periodic auctions to ensure market liquidity.

There are no existing U.S. models of an emissions tax, although five European countries have carbon-based taxes. [5] The closest U.S. example is the tax on ozone- depleting chemicals (ODCs). To facilitate the phaseout of ODCs under the Montreal Protocol and subsequent amendments, the United States imposed a tax on specific ODCs in 1990. This tax was designed to supplement the allowance trading program that the EPA had designed to implement the international agreements. Several activities trigger the tax, including the production and/or importation of the chemicals, or the importation of products that contain them or used them in their production processes. In addition, inventories of certain ODCs held on January 1 of each year are subjected to a "floor stocks tax." [6]

FIVE DIMENSIONS OF THE COST ISSUE

Five dimensions of costs associated with reducing GHG emissions are discussed in this section: (1) absolute costs, (2) distribution of costs, (3) long-term costs, (4) price signal and stability, and (5) uncertainty of costs.

The absolute costs of a GHG reduction program are a function of the interplay among the tonnage reduction required, the timetable imposed on that reduction, and the techniques available and used to achieve that reduction (the "three Ts"). Variables involved with the tonnage requirement include the magnitude and firmness of the reduction requirement and the number of gases and sectors involved in the program. Variables involved with the timetable include its length and number of phases, along with the number and extent of any deadline extensions allowed and on what basis. Finally, variables involved with techniques include promotion and availability of new technology, the degree of flexibility permitted in complying with the program, and any ceiling on compliance costs. All these program design parameters influence the absolute cost of the program and the timing and extent of any benefits received.

A second concern with costs is their distribution across the various sectors of the economy. As indicated by Table 1, GHG emissions are spread throughout the economy, with about 81% emitted by the electric power, transportation, and industry sectors. Restricting participation by any group could increase the absolute cost of the program and would certainly increase the costs to the remaining participants. However, numerous rationales have been put forward to justify excluding one group or sector from a reduction requirement, or to provide some other special consideration. Rationales offered include a sector or industry's concern about (1) international competitiveness, (2) lack of cost-effective control options, (3) inability to make necessary capital investments, (4) economic disruption, (5) credit for previous efforts that reduced emissions, and (6) the "minor" contribution that industry or sector makes to the overall problem. It is the multitude of such variables that make constructing an acceptable reduction allocation scheme very difficult.

Table 1. 2005 U.S. Greenhouse Gas Emissions

Economic Sector	Million Metric Tons of CO2 equivalent	Percentage of Total
Electric power industry	2,430	33.7%
Transportation	2,009	27.9%
Industry	1,353	18.8%
Agriculture	595	8.3%
Commercial	431	6.0%
Residential	381	5.3%
Total	7,199	100.0%

Source: U.S. Environmental Protection Agency, Inventory of U.S. Greenhouse Gas Emissions and Sinks: 1990-2005, EPA 430-R-07-002 (Washington, DC), p. ES-14
Note: The total does not include 62 million metric tons from U.S. territories.

A third concern is the long-term cost considerations of a GHG reduction program. Climate change policy has to be thought of in decades, not years. Ultimately, a successful climate change program would involve a long-term transition to a less carbon-emitting economy. Generally, studies that indicate the availability and cost-effectiveness of emerging new technologies to achieve this transition include an economic mechanism to provide the necessary long-term price signal to direct research, development, demonstration, and deployment efforts. [7] Developing such a price signal involves variables such as the magnitude and nature of the market signal, and the timing, direction, and duration of it. In addition, studies indicate combining a sustained price signal with public support for research and development efforts is the most effective long-term strategy for encouraging development of new technology. [8] As stated by Morgenstern: "The key to a long term research and development strategy is both a rising carbon price, and some form of government supported research program to compensate for market imperfections." [9]

A fourth consideration is the stability of the price signal in whatever form it takes (e.g., allowance prices, carbon taxes, auction prices). A stable and reliable signal is necessary to minimize economic disruption and to encourage new technology. Experience with existing emissions markets suggests that short-term price spikes and troughs occur that have at least short-term economic effects, either disrupting the market (in the case of high prices) or discouraging new technology (in the case of low prices). Causes of this volatility can include (1) lack of trading volume, (2) illiquidity in the market, (3) external events, and (4) regulatory uncertainty. History with previous emissions trading programs suggests that if a greenhouse gas program is based on a market-based implementation strategy, the inclusion of flexibility mechanisms to ensure reasonable market stability is desirable. [10]

A final cost consideration is the cost uncertainty presented by the wide range of projected costs of GHG reduction. To the extent one understands the variables that create the range presented by different forecasting models, one can design a program to address those variables. Projected costs of a proposed greenhouse gas reduction program will differ among models, based on the various economic and technological assumptions either embedded in the particular model's processes (endogenous variables) or assumed externally and inserted into the model. Weyant has identified five assumptions that explain many of the differences in greenhouse gas reduction program cost estimates [11]:

- Basecase projections of future GHG emissions and climate damages.
- The specifics of the reduction program examined (particularly the amount of flexibility permitted in complying with its mandates).
- How dynamic the model is, representing substitution possibilities by producers and consumers, including the turnover of capital equipment.
- How the rate and processes of technological change are modeled.
- How benefits are modeled.

Figure 1, below, illustrates how these and other variables (such as type of model used) can influence the estimated costs of climate change legislation. Measured by impact on GDP, the figure indicates impacts generally ranging from a positive 2% increase in GDP to a 4% decrease. Interestingly, the variables used in projecting cost and benefits are sufficiently robust to obscure a strong correlation between cost and reduction requirements.

The range indicated also reflects the perspectives and parameters assumed by the forecast authors. In a previous report, CRS noted that cost analyses are influenced by the perspective (or lens) through which one views the problem. [12] Analysts viewing climate change policy through a technological perspective see it as an impetus for improved efficiency through technology improvements in the economy, consistent with concepts such as life-cycle costs. Analysts viewing policy through an economic lens work through the boundaries of market economics and cost-benefit considerations. Finally, analysts viewing the issue through an ecological lens look to the benefits of controlling greenhouse gases and are suspicious of "baseline" scenarios that suggest that "business as usual" is an acceptable yardstick from which to measure policy changes. Each of these lenses implies fundamentally different ways of assessing policy actions and modeling potential costs and benefits. The quantitative results are cost estimates that range from actual savings to the economy (from GHG reductions) to substantial costs.

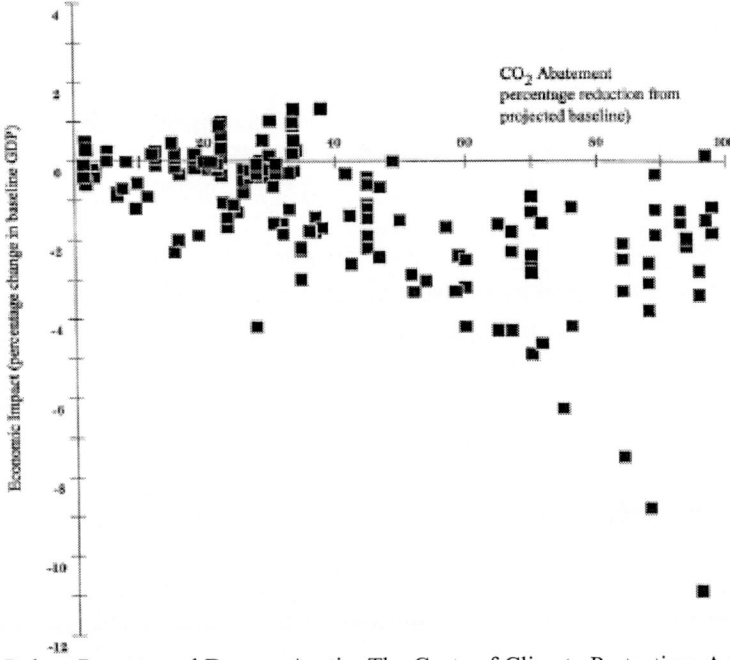

Source: Robert Repetto and Duncan Austin, The Costs of Climate Protection: A Guide for the Perplexed (World Resources Institute, 1997), p. 12.

Figure 1. The Predicted Impacts of Carbon Abatement on the S. Economy (162 Estimates from 16 Models).

APPROACHES FOR ADDRESSING COST CONCERNS

The following analysis of options to address the cost concerns identified above is loosely arranged by the focus of the specific option: (1) the tonnage requirement, (2) the time frame, and (3) the techniques allowed for compliance. It should be noted that several options examined affect more than one of the "Ts." Also, the options are not mutually exclusive — many can be combined to create more refined options.

Tonnage Options

Much of the discussion on GHG reductions has focused on a historic baseline as the starting point for reductions. Assuming that the emissions inventory for a specific year is adequate to support a regulatory program (whether market-based or not), such a baseline is reasonable. Most existing emissions trading programs are based on a historic baseline with modifications. However, there are options to calculate a baseline that responds to economic events over time without necessarily compromising the tonnage cap. Also, the historic baseline can be eliminated in favor of different methods of achieving specific reductions.

Dynamic Tonnage Target. Another approach to address some of the concerns identified above is to calculate the tonnage target based on economic or other indexes or measures rather than strictly on a historic or other static baseline. For example, the National Commission on Energy Policy recommended that the tonnage requirements for a GHG reduction program begin with year 2000 emissions, with the future trajectory of emissions based on the product of a progressively declining limit on the country's GHG intensity times projected economic growth. Over time, the progressively more stringent carbon intensity index would produce progressively more stringent emissions tonnage caps, despite projected increases in economic growth. The actual steepness of that path would depend on the rate of decline in carbon intensity mandated by the program and actual economic growth. Of course, the dynamic tonnage target could be indexed to just about any relative variable (e.g., energy prices).

Depending on the specifics of the methodology and measures used in creating it, the dynamic target could be more responsive to some unforeseen events, such as substantial economic growth, than a static baseline. At least in the short term, this could reduce costs and economic disruption if a sharp spike in economic growth were to occur. In contrast, slower-than-anticipated growth would reduce the available emissions credits and thereby reduce the potential for "hot air" credits (i.e., credits "created" by a slowdown in the economy rather than by control efforts). [13]

By potentially mitigating some effects of a static, historic emissions baseline, a dynamic tonnage methodology allows flexibility in distributing reductions and the resulting costs among different sectors of the economy. Growth, GHG intensity, production, and other variables could be tailored for sectors, states, or regions based on specific concerns, such as competitiveness. For example, an industry growth index could be used to calculate reduction requirements rather than an aggregated index such as GDP. Like most schemes, a dynamic target scheme could completely exclude some industries, with the obvious result of a shift in cost to the ones remaining in the program.

The effect of a dynamic target on long-term costs would depend on the slope of reductions mandated by the program. For example, the recommendation of the National Commission on Energy Policy called for an annual 2.4% reduction in allowable GHG intensity increasing to 2.8% annually after 10 years. This declining curve would be multiplied by a projection of presumably increasing economic growth. A steeper slope in GHG intensity mandates and/or an overly pessimistic projection of economic growth would strengthen the need for less carbon-intensive technology, but at the risk of increasing cost if those technologies did not arrive in a timely manner. A weak GHG intensity mandate and/or an overly optimistic projection of economic growth could reduce necessary emissions reductions and provide a weak incentive for new technology.

A dynamic target would not necessarily prevent short-term fluctuations in the price signal, depending on the frequency of adjustments. If a target were based on macro-economic trends, such as GDP, it would not respond much to short-term or localized events, such as the 2000-200 1 electricity shortage in California. Also, the mixture of indices with different vectors (e.g., GHG intensity reducing targets while economic growth is increasing targets) may create some uncertainty in markets regarding the appropriate price of credits.

Finally, a dynamic target would not increase the certainty of cost estimates. Uncertainty about the future trajectory of economic growth would be reflected in cost estimates (just as they are now, with emissions capped at historically determined levels). Likewise, benefit certainty would not improve for the same reason.

Expand Supply Options. The breadth of options permitted under a reduction program can have a significant effect on absolute costs. Legislation introduced in recent Congresses has ranged from programs based on one economic sector (e.g., electric utilities) and one greenhouse gas (e.g., carbon dioxide) to several sectors (including opt-in provisions) and all six greenhouse gases covered by the Kyoto

Protocol. [14] Also, some proposed programs have included international trading of emissions credits and biological sequestration offsets among the permissible means of complying with reduction requirements. [15] Some of these options, particularly international trading and sequestration, have included limits on their applicability. For example, the Regional Greenhouse Gas Initiative [16] has put control cost triggers (characterized as "safety valves") on the availability of some supply options, such as sequestration.

Numerous analyses were done on the impact of global trading after the signing of the 1997 Kyoto Protocol. For the United States, the cost of complying with the Kyoto Protocol was estimated at $23-$50 per ton of carbon if global trading were included, versus $61-$119 if only trading among developed (Annex 1) countries were permitted. Cost estimates of "No trading" scenarios ranged from $193-$295 per ton. [17] Studies have suggested that, beyond international trading, including non-carbon dioxide greenhouse gases and sequestration in the supply mix can play an important and cost-effective role in any climate change program. [18]

Expanding supply sources could help industries that do not have readily accessible means of reducing greenhouse gases on their own by providing them with additional options and making the credit market more liquid. To the extent the expanded supply sources help create an integrated market with a true market price for credits, industries could avoid very high compliance costs and lessen the impact of those costs on their profitability. However, if competitors in other countries do not have to reduce emissions at all (as is currently the case with the Kyoto Protocol), competitive disadvantage would remain in some cases.

The degree to which expanded supply options would contribute to a long-term and stable price signal would depend on how integrated these sources are in the overall permit market. For example, with the European Trading System (ETS), there are separate markets for credits created within the 15 members of the European Union (EU) covered by the EU bubble, credits created by Joint Implementation with eastern European countries, and credits created via the Clean Development Mechanism (CDM) with Third World countries. [19] One type of credit cannot be traded for another. The result is a range of credit prices, reflecting the relative risk and availability of the various credit types. Thus, the long-term signal being delivered is currently unclear, and may take time to develop. Likewise, substantial fluctuations in the EU credit market have not been stabilized by the existence of the other two credit types.

In some ways, expanding supply options may increase the uncertainty of cost estimates, not only because of disparity in assumed reduction costs, but also in assumed availability and penetration of the options themselves. For example, emissions reductions via the Clean Development Mechanism could be substantial and very cost-effective. However, the mechanism itself creates uncertainty with respect to availability, as does the willingness of foreign governments to participate. It is difficult to quantify the effect such an option could have on costs without some track record, as is slowly being built by the ETS.

Carbon Tax. The most radical approach to controlling costs and addressing the concerns identified above is to impose a carbon tax in lieu of proposed allowance trading programs. As discussed in the introduction to this report, under a carbon tax, the costs are fixed by the legislation and the quantity of emissions reduced becomes the variable. Carbon taxes are generally conceived as a levy on natural gas, petroleum, and coal, according to their carbon content, in the approximate ratio of 0.6 to 0.8 to 1.0, respectively. However, the levy would not have to be imposed on the fuels themselves; proposals have been made to impose the tax downstream at the point where the fuel is converted into heat and CO2. In addition, there is no reason why the tax could not be expanded to include all greenhouse gases in appropriate carbon equivalents.

A carefully designed carbon tax could potentially address all five of the concerns identified above. A carbon tax puts a limit on absolute cost by capping the marginal costs that participants should pay to reduce GHG emissions. Participants would receive a firm price signal with respect to the upper value of GHG emissions, and respond in the most cost-effective manner — that is, reduce emissions up to the cost of the carbon tax and pay the tax on any remaining emissions that are more expensive to eliminate.

A carbon tax can be tailored to address distributional concern in two ways. The first would be to exempt, either partly or completely, whatever sectors or industries were felt to be threatened, either competitively or otherwise, by imposing the tax. The current tax code provides numerous exemptions from various taxes for a variety of reasons. However, such an approach would create economic distortions and complicate the tax structure. The second approach would be to use some of the revenue generated by the tax to provide appropriate relief to targeted sectors or industries. This could involve increasing funding for existing programs for such sectors or industries, or creating new ones. In some ways, this approach might be more transparent than an approach that involves a potentially complicated tax structure. These approaches are not mutually exclusive; they could be combined if considered appropriate.

Likewise, a carbon tax can be employed to address long-term concerns in two ways. First, the carbon tax would create a long-term price signal to stimulate innovation and development of new technology. This price signal could be strengthened if the carbon tax were escalated over the long run, either by a statutorily determined percentage or by an index (such as the producer price index). Second, some of the revenue generated by the tax could be used to fund research, development, demonstration, and deployment of new technology to encourage the long-term transition to a less-carbon-intensive economy.

A carbon tax's basic approach to controlling GHG emissions is to supply the marketplace with a stable, consistent price signal — the fourth cost concern. Designed appropriately, there would be little danger of the price spikes or market volatility that can occur in the early stages of a tradeable permit program.

Finally, a carbon tax basically places an upper boundary on projected economic cost uncertainty. However, it increases uncertainty with respect to environmental benefits by making emissions reductions a dependent variable. This is the basic tradeoff that a price-based control system presents. One way that might mitigate the problem to some extent would be to combine the carbon tax with some form of quantity controls. As noted earlier, the CFC program attached a tax to its trading program with beneficial results. However, it is the trading program, not the CFC tax, that is the primary regime for control. In this manner, a carbon tax would be more of a revenue raiser than a control regime. A second hybrid would be a "safety valve" that capped allowance prices such as proposed by the National Commission on Energy Policy. [20] That approach is discussed later in this report. The degree to which the problem is mitigated (and others created) depends on the interplay between the quantity control and the carbon tax.

Timetable Options

Similar to the country's three-plus decades effort to reduce smog, climate change promises to be an effort measured in decades, not years. Unlike conventional pollution control efforts, the environmental benefit of mitigating climate change would come from a reduction in the stock of greenhouse gases that have built up in the atmosphere for decades, whereas the economic costs of control are related to the current flow of additional gases into the atmosphere. Thus, in a situation similar to protecting the stratospheric ozone layer, there would be a substantial delay between control costs and environmental benefits. Indeed, if short-term reductions in the stock of greenhouse gases were the focus of climate change policy, control efforts would be focused on controlling methane, which has a 20-year lifetime, compared with CO2, which has a 200-year lifetime. Likewise, temporary measures, such as biologic sequestration, would be accelerated with the assumption that new technology would be available in the future to capture the biologically sequestered carbon dioxide when it is released decades from now.

This situation leads to disputes over how time should be managed under a GHG reduction program. One argument is that modest cuts (or slowing of the increase) early, followed by steeper cuts later, is the most cost-effective. Generally, three cost- related arguments are made in favor of this approach. First, over the long-term, sustained GHG reductions involve a turnover in existing durable capital stock — a costly process. If the time frame of the reduction is long enough to permit that capital stock to be replaced as it wears out, the transitional costs are reduced. Second, increased time to comply would permit the development and deployment of new, less carbon-intensive technologies that are more cost-effective than existing technology. Third, assuming a positive rate of return on current investment, less money needs to be set aside today to meet those future compliance costs. [21]

A counter argument to the above focuses on the risks of delay, both in terms of scientific uncertainty and technology development. In terms of scientific uncertainty, there is no consensus on what concentration of greenhouse gases should not be exceeded in order to avoid undesirable climate change. If the stabilization level needed is relatively low, any delay in beginning reductions could be costly, both economically and environmentally. [22] Secondly, given the sometimes long lead times for technology development, both a long-term price signal and research and development funding may have to be initiated quickly to encourage technology development and deployment in time to hold GHG concentrations to a level that avoids unacceptable damages. In the same vein, an early signal with respect to climate change policy is necessary to discourage investment in durable long-lived (50-60 years) carbon-intensive technologies. [23] As stated by Jaccard and Montgomery:

> The window of opportunity for reducing cost implies a need for immediate and continuing action to develop new low-carbon technologies and to begin shifting long-lived investment decisions toward alternatives that lower carbon emissions. Absent these actions, the rapid future emissions reductions included in the delayed emissions scenario may be more costly than more evenly paced, and earlier reductions. [24]

Economic-Based Circuit Breaker. Delaying or suspending compliance with environmental mandates because of energy and economic reasons is not a novel idea. The Clean Air Act contains provisions permitting the President, in response to a petition by an affected state's governor, to temporarily suspend any part of a state implementation plan or enforcement of the SO2 trading program, to address a severe national or regional energy emergency. [25] For example, during the 2000-200 1 California energy crisis, President Clinton directed all federal agencies to do their part to assist the state in meeting its electricity demand. For its part, the EPA revised its guidance on emergency generators to allow backup generators to be used to avert a power blackout. [26] Previously, backup generators could be used only when the power was actually interrupted. The increased flexibility permitted by the EPA during the emergency meant more power at the expense of more pollution (particularly of carbon monoxide and nitrogen oxides).

Likewise, market-based systems are not immune to being suspended if economic or energy conditions turn severe. A contributing factor in the California power crisis was the Regional Clean Air Incentives Market (RECLAIM), a credit trading system for reducing nitrogen oxide (NOx) emissions. RECLAIM was established in 1994 to provide flexibility for companies in the South Coast (Los Angeles) area as controls on NOx, a major contributor to smog formation, were tightened. Because of record electricity demand in 2000, electric generators in the South Coast area generated more power than they did in the base period, resulting in utilities buying RECLAIM trading credits in unprecedented quantities. As a result, the price of credits rose from less than $1 per pound of NOx in January 2000 to more than $60 per pound of NOx by March 2001. [27] To solve this problem, in March 2001, the South Coast Air Quality Management District amended RECLAIM to remove large power plants from the trading system and required owners of such facilities to reduce emissions under a mandated command-and-control regime. Such facilities returned to the trading system in 2007. [28]

Proposals have been made to formalize a "circuit breaker" into any GHG reduction program. In general, proposals envision a declining emissions cap system where the rate of decline over time is determined by the market price of permits. If permit prices remain under set threshold prices, the next reduction in the emissions cap is implemented. If not, the cap is held at the current level until prices decline. [29] Such a cap could be implemented on an economy-wide basis or by sector or other relevant grouping.

Because the conditional reduction approach attempts to turn both the price and the quantity of reductions into variables solved by the trading market, its effect on cost depends on a host of variables — most obviously the profiles of the emissions reduction targets and the price triggers. For example, the price trigger could be based on the spot-market price, the long-term market price, or some hybrid price mechanism. Also, the reliance on the market to either directly or indirectly determine price and quantity puts pressure on regulators to oversee operations and prevent any market manipulation designed to slow emissions reductions.

A conditional tonnage target could address distributional issues if its tonnage targets and timetable triggers are tailored for specific sectors or industries. This would substantially increase the complexity of the scheme and potentially risk bifurcation of the permit market. As with other permit schemes discussed here, another approach to addressing distributional concerns under a conditional tonnage target would be to simply exempt certain sectors from its mandates.

When the control regime responds to relatively short-term events, it may not provide the long-term price signal necessary to promote long-term solutions. The elastic time frame also gives ambiguous signals for planning the appropriate pace and scope of research and development efforts. In contrast, the regime's focus on short- term economic disruption may help in damping short-term volatility in the allowance market. As noted, the responsiveness of the price and timetable triggers would determine how effective the program would be in avoiding such disruption.

In reducing the uncertainty for cost estimates, the scheme introduces about the same number of new uncertainties as it does in reducing others. The circuit breaker prevents ever increasing costs, although with some undetermined lag time. However, it is difficult to estimate absolute costs because one does not know how often it will be used.

The scheme also increases uncertainty about the future trend in benefits by making the quantity of emissions reduced a variable. A short-term break might not make much of a difference, particularly if participants were required to make up the emissions later. However, it introduces uncertainty into the system that would be difficult to quantify.

Technology-Based Timetable. Another approach to increase flexibility in the system and encourage long-term technology development would be to provide special compliance schedules for entities deploying innovative, less carbon-intensive technologies. An example of this possibility is Section 409 of Title IV of the 1990 Clean Air Act Amendments. [30] Under Section 409, utilities choosing to meet their sulfur dioxide reduction requirements by installing a qualifying clean coal technology receive a four-year extension on the program compliance deadline. During the extension, the affected emitter is allowed to operate under existing regulations and operating conditions. If the technology fails to operate as designed, the affected unit may be retrofitted with another qualifying technology or with an existing control technology. For a GHG reduction program, a qualifying technology could include geologic sequestration, emerging energy efficient technology, or advanced solar power.

A technology extension could reduce costs in two ways. First, the delay in compliance itself would reduce cost by allowing the affected company more time to gather resources and optimize a compliance plan. Second, to the extent the delay encourages more cost-effective approaches to GHG reductions, compliance cost and long-term cost would be reduced. Of course, the risk is that the delay will not result in successful technology development. Indeed, it is likely that at least some of the projects would fail — that is the nature of innovation. However, because technology development is crucial to long-term reductions in greenhouse gases, some may feel the risk is worth it.

Assistance with distributional costs under this option would depend on the opportunities for new technology in given sectors of the economy. Although some industries may have potentially cost-effective technology-fixes, such as geologic sequestration, others may involve long-term structural changes.

Of course, the focus of a technology-based timetable is to provide a long-term signal to the market encouraging new technology. Such a signal could be strengthened significantly with increased government funding of projects.

Because this option is focused on new technology, it would seem likely to have little effect on short-term price volatility. However, there may be a risk that the temporary removal of significant emitters from the market-system in response to the incentive could increase short-term volatility and uncertainty by diminishing permit demand and trading volume.

It is difficult to determine the effects of this option on cost estimates. It would depend on how widespread the assumed participation rate is.

Technique Options

Most current GHG reduction proposals assume a market-based implementation strategy — generally a permit trading program. This is not surprising, as flexibility and new technologies are considered the keys to a cost-effective implementation strategy over the long run. Generally, technique options range from making a tradeable permit program more flexible through mechanisms like banking, to creating a hybrid program where the regime shifts from a quantity-based permit program to a carbon tax, depending on defined circumstances.

Banking and Borrowing. Most existing trading programs include provisions for banking credits for either future use or future sale. Indeed, the absence of effective banking in the RECLAIM program (discussed earlier) is credited with contributing to RECLAIM's suspension during the California energy crisis. As summarized by Resources for the Future (RFF):

> Allowance banking has been an essential component of the SO2 program. Its absence is a costly feature of the NOx programs, eroding the opportunity for cost savings from interannual trading and contributing directly to the suspension of trading in RECLAIM. [31]

Banking and borrowing reduces the absolute cost of compliance by making annual emissions caps flexible over time. The limited ability to shift the reduction requirement across time allows affected entities to better accommodate corporate planning for capital turnover and technological progress, to control equipment construction schedules, and to respond to transient events such as weather and economic shocks. Generally, banking and borrowing would not have any direct impact on distributional concerns, which are more directly determined by initial allocation decisions. Banking and borrowing can help provide a long-term market signal by supporting credit prices when costs are lower than expected. [32]

The flexibility provided by banking and borrowing, as noted, can help dampen short-term volatility. The degree that they help is disputed. As discussed later, some argue that banking and borrowing may provide sufficient flexibility in some cases to keep market disruptions to a minimum. [33] However, others argue that if a program involves more than modest reductions, a more robust "safety valve" is preferable. [34]

In estimating costs, banking and borrowing help smooth out the reduction requirement, as witnessed by the current acid rain program. This economically desirable effect does not necessarily reduce the uncertainty in cost estimates because estimators will make different assumptions about the extent to which banking and borrowing are used by emitters. The smoothing effect, however, has no effect on the reduction requirement (in contrast with several of the other alternatives discussed here). This is a major reason why this alternative is generally favored by those whose priority is to achieve specific reductions.

Auctioning Permits. Auctions can be used in market-based pollution control schemes in several different ways. For example, Title IV of the 1990 Clean Air Act Amendments uses an annual auction to ensure the liquidity of the credit trading program. For this purpose, a small percentage of the credits permitted under the program are auctioned annually, with the proceeds returned to the entities that would have otherwise received them. Private parties are also allowed to participate. A second possibility is to use an auction to raise revenues for a related (or unrelated) program. For example, many states participating in the

Regional Greenhouse Gas Initiative (RGGI) are using auctions to implement their public benefit programs to assist consumers or pursue strategic energy purposes. [35] A third possibility is to use auctions as a means of allocating some, or all, of the credits mandated under a GHG control program. In examining a modified auction program, a Resources for the Future (RFF) analysis found that an auction scheme is "dramatically more cost- effective" in allocating credits than either a grandfathered allocation method [36] or a generation performance standard [37] (GPS) approach. [38] Obviously, the impact that an auction would have on the cost dimensions identified earlier would depend on how extensively it was used in any GHG control program, and to what purpose the revenues were expended.

The cost-effectiveness of an auctioning system results from allowing the marketplace to allocate credits. However, unlike a carbon tax, the market-clearing price for credits is not limited (unless the system is combined with a safety valve and/or a reserve price, as discussed below). Hence, an auction for credits would be more expensive for specific industries than under a historically based grandfathered system, where they would receive their credits free. Likewise, the price consumers pay may be greater, depending on the companies' ability to pass on their additional costs to them. However, when the substantial revenues received by the auctions are considered, auctions are more cost-effective than grandfathered or GPS systems. As stated by RFF:

> The bottom line is that the AU [auction] approach weighs in at substantially less economic cost to society than either of the two gratis approaches to allocating allowances.... AU also provides policymakers with flexibility, through the collection of revenues that can be used to meet distributional goals or to enhance the efficiency of the AU even further by reducing pre-existing taxes. Because the AU approach is so cost-effective, a corresponding a [sic] carbon policy will have less effect on economic growth than under the other approaches. This attribute provides the most significant form of distributional benefit. [39]

As noted by RFF, the revenues from an auction can be used to address a host of distributional concerns. Indeed, as noted earlier, the auction could be tailored to raise only as much as necessary to address those concerns (as with RGGI funding of public benefits programs) or made more comprehensive to address credit allocation.

In terms of a long-term price signal, the type of auction employed would have some effect. For example, the program could implement a price floor to facilitate investment in new technology via a reserve price in the allowance auction process. In addition, the stability of that price signal could be strengthened by choosing to auction allowances on a frequent basis, ensuring availability of allowances close to the time of expected demand and making any potential short-squeezing of the secondary market more difficult. [40]

An auction could provide substantial incentive for new technology if the auction is structured to encourage a long-term and stable price signal and if revenues received are at least partly directed toward research, development, and demonstration programs.

An auction would probably not reduce the uncertainty with costs, because differing assumptions could be made about the actual operation of the auction, its efficiency, and the effectiveness of the recycled revenues. However, an auction would not have any effect on benefits received by the program, unless it were joined with a safety valve or other limit on auction prices.

Safety Valve. The purpose of a safety valve is to limit the costs of any climate change control program (price) at the potential expense of reductions achieved (quantity). Safety valves encompass a variety of carbon tax-tradeable permit hybrid schemes. Perhaps the most publicized version is that recommended by the National Commission on Energy Policy. [41] The Commission scheme would be implemented through a flexible, market-oriented permit trading program. The total number of permits each year would be based on a mandated decline in GHG intensity and projected GDP growth. However, the scheme includes a cost-limiting safety valve that allows covered entities to make a payment to the government in lieu of reducing emissions. The initial price of such payments would be $7 per ton in 2010. Thus, if a covered entity chooses, it may make payments to the government at a specific price rather than make any necessary emissions reductions.

Effectively, a safety valve places a ceiling on compliance costs; in that way, it acts like a carbon tax. To the extent an entity's control costs, or the permit market, remain below the safety valve, the scheme acts like a tradeable permit program. The degree to which a safety valve reduces costs would depend on the extent to which it is used by entities (e.g., who do not have a cost-effective alternative). However, the complex interactions involved in a scheme that includes both price and quantity controls should not be underestimated. As stated by Jacoby and Ellerman:

> The usefulness of the safety valve depends on the conditions under which it might be introduced. For a time, it might tame an overly stringent emissions target. It also can help control the price volatility during the introduction of gradually tightening one, although permit banking can ultimately serve the same function. It is unlikely to serve as a long-term feature of a cap-and-trade system, however, because of the complexity of coordinating price and quantity

instruments and because it will interfere with the development of systems of international emissions trade. [42]

In contrast, Morgenstern argues that the complexity is worth it in preventing price spikes, particularly if a substantial reduction in emissions is envisioned: "If only modest reductions are undertaken, a system of banking and offsets is likely to be adequate in preventing price spikes. In order to achieve more ambitious targets, however, the safety valve is clearly preferred." [43]

To address distributional concerns, a safety value could be tailored for specific sectors to address concerns about cost-effective reduction options or competition. In addition, to the extent the safety valve created revenues, some of the funds raised could be recycled to affected parties.

The effect of a safety valve on new technologies reflects the complexity discussion above. If a low safety valve price were chosen (meaning it would keep compliance costs low), it could have a dampening effect on long-term development of new technology. By creating a ceiling on the value of GHG reductions, but providing no floor for those reductions, a weak market signal may be sent. This might be offset to some degree if funds collected by the safety valve were directed toward new technology, but marketing of any resulting technology might still be difficult if the market price is held low.

A safety valve would dampen the possibility of an upward spike in credit prices — indeed, it is a major reason for considering such an option. However, it would not affect any volatility occurring below the safety valve value and have no effect on a collapse in credit prices. By the same token, the safety valve would put an absolute limit on the projected costs of the program at the level of the safety valve. However, it would do this at the expense of certainty in terms of reductions achieved.

ILLUSTRATIVE APPROACHES

The selected options discussed above are summarized in Appendix A. As suggested, the various options identified have different strengths and weaknesses, depending on the facet of costs one wants to address. Fortunately, many of the options are not mutually exclusive, nor do they require complete adoption; parts of individual approaches can be combined with other parts to meet program specifications in terms of firmness of the goal (also called the "hardness" of the emissions cap) and time frame.

To illustrate, a program focused on achieving a specific tonnage reduction with some flexibility in implementation but not in a manner that threatens the integrity of the cap could incorporate several of these options. The most obvious mechanism to include in the quantity-based cap-and-trade would be banking and borrowing options that would increase flexibility of the program across time without any deterioration in the tonnage requirement. Flexibility and protection against price increases could be enhanced by expanding supply options to include all greenhouse gases, sequestration, and international trading. Depending on one's confidence in the individual supply options, use could be restricted to a maximum percentage of reduction achieved through the option (common in many proposals) or to a more flexible percentage restraint based on credit prices (as in RGGI's cap-and-trade scheme). Proper monitoring and enforcement could minimize any potential effect on the cap.

This illustration would not necessarily provide either the long-term price signal or funding necessary for new technology. One supplemental option that could help mitigate this problem would be credit auctions. Auctions would have no effect on the cap, but would provide the program with a revenue flow that could be at least partly directed toward research and development. The auction could be designed to raise revenues only by auctioning a small percentage of the credits (such as the current acid rain program), or be comprehensive and auction all credits, thus improving overall economics and providing a clear market signal (as many RGGI participants are doing). In the latter case, coordinating the auction with any trigger price mechanism for expanded supply options would promote harmonious implementation. Depending on the structure of the auction chosen, the comprehensive auction would also provide a clear market price for reductions and, with the addition of forward markets, some indication of the general direction of those prices.

Finally, the auction and its resulting revenues could also be used to address pressing distributional cost issues. Although the mixture of options used in this illustration could potentially mitigate several of the cost issues identified here, it would not provide cost certainty. The quantity side of the equation is the controlling factor under this illustration; prices could be tempered by the market flexibility introduced by the options, but actual costs would not capped.

In contrast, a more price-oriented illustration could employ a safety valve to place an absolute limit on credit prices. In such a hybrid system, the focus of the program is the safety valve limit as much as any tonnage cap. The quantity-based limits of the emissions cap determine the probability that the safety valve would be triggered, assuming a well-functioning market. However, in addition to the supply- demand dynamic that the credit market will reflect, any market failure or disruption resulting from external events could trigger the safety valve for participants. Ultimately, quantity is subordinate to price.

One can potentially reduce the probability that the safety valve would be invoked by including several of the other options discussed here. Expanding supply options would enlarge the pool of available reductions and potentially improve the stability of the credit market if properly integrated. Employing a dynamic tonnage target or an economic-based circuit breaker could help address any economic growth spike that might trigger the safety valve. The question of using these options in a safety valve program is whether they would affect the cap more or less than invoking the safety valve. In contrast, borrowing and banking would help stabilize markets without having any effect on the cap.

Like the illustration above, this approach would not necessarily promote new technology — indeed the safety valve could discourage such development, unless it generated revenue that was directed toward research and development. If revenues were deemed insufficient for new technology (and to address distributional concerns if desired), the safety valve program could be supplemented with an auction. However, in any case, this illustration is driven by price concerns — concerns that make coordinating new technology development and minimizing impacts on the emissions cap difficult.

A final illustration could also be the simplest — imposition of a carbon tax. The clear focus of the program would be the level of the tax, the steepness of any future increases in the tax, and who has to pay the tax. As noted earlier, it could be crafted to address all the cost concerns identified in this report; however, it would represent a new direction in U.S. climate change and current international efforts.

ADDRESSING COSTS THROUGH MARKET MECHANISMS: RESOLVING THE PRICE-QUANTITY ISSUE

The fundamental policy assumption that changed between the U.S. ratification of the 1992 United Nations Framework Convention on Climate Change (UNFCCC) and the George W. Bush Administration's 2001 decision to abandon the Kyoto Protocol process concerned costs. [44] The ratification of the UNFCCC was based at least partially on the premise that significant reductions could be achieved at little or no cost. This assumption helped to reduce concern some had that the treaty could have deleterious effects on U.S. competitiveness — a significant consideration because developing countries are treated differently from developed countries under the UNFCCC. Further ameliorating this concern, compliance with the treaty was voluntary. While the United States could "aim" to reduce its emissions in line with the UNFCCC goal, if the effort indeed involved substantial costs, the United States could fail to reach the goal (as has happened) without incurring any penalty under the treaty. This flexibility would have been eliminated if the United States had ratified the Kyoto Protocol with its mandatory reduction requirements; the George W. Bush Administration cited this lack of flexibility as a major reason for rejecting the Kyoto process.

Two events provide impetus for revisiting the cost issue with respect to designing a greenhouse gas reduction program. The first is the election of a new President publicly committed to substantial reductions in greenhouse gases over the next several decades. The second was passage of the 2005 Sense of the Senate climate change resolution calling on Congress to enact a mandatory, market-based program to slow, stop, and reverse the growth of greenhouse gases, and which states that the program should be enacted at a rate and in a manner that "will not significantly harm the United States economy" and "will encourage comparable action" by other nations. [45] Facets of the cost issue that have raised concern include absolute costs to the economy, distribution of costs across industries, competitive impact domestically and internationally, incentives for new technology, and uncertainty about possible costs.

Market-based mechanisms attempt to address the cost issue by introducing flexibility into the implementation process. The cornerstone of that flexibility is permitting sources to decide their appropriate implementation strategy within the parameters of market signals and other incentives. That signal can be as simple as a carbon tax or comprehensive credit auction that tells the emitter the value of any reduction in greenhouse gases, to a credit marketplace that is constrained by a ceiling price (safety valve) and includes incentives for new technology. As illustrated here, the combinations of market mechanisms are numerous, allowing decision makers to tailor the program to address specific concerns.

In a sense, the options discussed here represent a continuum between alternatives focused on the price side of the equation (e.g., carbon taxes) through hybrid schemes (e.g., safety valves) to alternatives focused on the quantity side (e.g., banking and borrowing). They are tools to assist in the assessment of potential greenhouse gas reduction approaches, leaving any policy decision on balancing the price-quantity issue to the ultimate decision makers.

Appendix A. Summary of Selected Options To Address Cost Uncertainty of Greenhouse Gas Reduction Programs

Option	Absolute Costs	Distributional Costs	Long-Term Costs	Price Stability	Cost Uncertainty	Effect on Benefits
Carbon Tax	Allows economics to determine ultimate emissions reductions. Costs limited to tax levy. Actual costs would depend on the level of the tax, availability of reduction below that level, and the distribution of the revenues.	Distributional concerns about costs can be addressed by either partly or completely exempting specific sectors or targeting sectors with funding from the received tax revenues.	Long-term development of new technology would be stimulated by creating a long-term price floor on carbon and strengthened further by targeting R&D with funding from the received tax revenues.	Would provide a stable, consistent price signal.	Would provide an upper limit on potential cost estimates. The lower limit would still be subject to uncertainty.	Would make reductions dependent on the level of the tax. The quantity reduction becomes the variable while the price is fixed.
Dynamic Tonnage Target	Depending on specifics, would probably offer some cost protection against unforeseen spikes upward in economic growth.	Distributional concerns about costs could be addressed by variety of regional or sector-specific metrics.	Incentive for new technology would depend on the slope of reductions mandated by the program.	Would not necessarily avoid short-term fluctuations in market price. Different metrics for different sectors could also create market price uncertainty.	Would only have modest effect on reducing uncertainty in cost estimates.	Depending on the specifics of the target, benefits could be at least slightly dependent on economic conditions.
Expanded Supply Options	Can substantially reduce costs, depending on the additional options included.	Can help sectors that do not have cost-effective means of reducing emissions on their own.	Depends on how well the additional sources are integrated into the overall market — a stratified market can muddle the long-term price signal.	Depends on how well the additional sources are integrated into the overall market — a stratified market may result in independent pricing trends.	Can increase uncertainty by adding new variables to the estimates, including availability, penetration, and costs of the additional options.	Should have no effect on reductions achieved, assuming proper safeguards are taken, but new risks are introduced with some options (like international trading).

Appendix A – Continued

Option	Absolute Costs	Distributional Costs	Long-Term Costs	Price Stability	Cost Uncertainty	Effect on Benefits
Economic-Based Circuit Breaker	Reduces costs by temporarily extending compliance deadlines and/or slowing emissions reduction targets. The degree of cost savings depends on the specifics of the program.	Could address distributional concerns by tailoring its tonnage and timetables triggers to specific sectors.	Its short-term focus could muddle the long-term price signal important for developing new technology.	Depending on the responsiveness of the tonnage and timetable triggers, it would help mitigate short-term price volatility.	Scheme introduces many new uncertainties while reducing others in estimating costs.	Increases uncertainty benefits by making the quantity of reductions achieved a variable.
Technology-Based Timetable	Potentially reduces costs by delaying compliance and encouraging more cost-effective approaches in the long-term.	Would depend on opportunities for new technologies in given sectors.	Arguably, the primary focus of this scheme is to encourage new technology deployment.	May have little effect on price stability; indeed, it could increase short-term volatility and uncertainty by removing demand and volume from the market.	Scheme introduces new uncertainty to cost estimates.	Effect would depend on how widespread the assumed participation rate is.
Banking and Borrowing	Reduces costs by making the emissions cap more flexible over time.	Little effect.	Can help support a long-term price signal for new technology by supporting prices when costs are lower than expected.	The added flexibility can help damp short-term price volatility, but not eliminate it.	No significant effect.	No significant effect over the long-term.

Appendix A – Continued

Option	Absolute Costs	Distributional Costs	Long-Term Costs	Price Stability	Cost Uncertainty	Effect on Benefits
Auctioning Permits	Allows the marketplace to allocate permits. Actual costs would depend on percentage of permits auctioned and distribution of the revenues.	Can be used to address concerns by tailoring auctions for specific sector and/or directing revenues toward affected sectors.	Some revenues could be targeted for new technology. Also, auctions would help determine market price of reductions.	Depending on the volume of the auction, could have some effect on short-term volatility, but not eliminate it.	Scheme introduces new uncertainties to cost estimates.	No significant effect.
Safety Valve	Effect on cost depends on level that the safety valve is set.	Safety valve levels could be tailored for specific sectors.	By setting a ceiling but not a floor on prices, could have a damping effect on new technology depending on the level imposed.	Would place an upper limit on price volatility.	Would place an upper limit on cost estimates.	Would make reductions a function of the safety valve level.

This balance will not be easy to achieve. By offering flexibility to program designers and participants, market-based mechanisms can assist in implementing a GHG reduction program at less cost than more traditional command-and-control methods. [46] However, the complexity of market mechanisms (particularly trading programs) increases substantially with the scope of emitting sources included (particularly if international trading is envisioned) and the specificity of any allocation scheme. Thus, perhaps the most difficult issue to be addressed in designing a market-based implementation strategy for reducing GHG emissions is determining who is included, and who is exempted.

REFERENCE

[1] For an analysis of federal policy and congressional debate since ratification of UNFCCC, see CRS Report RL30024, Global Climate Change Policy: Cost, Competitiveness and Comprehensiveness, by Larry B. Parker and John E. Blodgett.

[2] President George W. Bush, President Bush's Speech on Global Climate Change (June 11, 2001).

[3] The six gases recognized under the Kyoto Protocol are carbon dioxide (CO_2), methane (CH_4), nitrous oxide (N_2O), sulfur hexafluoride (SF_6), hydrofluorocarbons (HFC), and perfluorocarbons (PFC).

[4] S.Amdt. 866, passed by voice vote after a motion to table failed 43-54, June 22, 2005.

[5] Finland, the Netherlands, Sweden, Denmark, and Norway.

[6] For CFC-11 and 12, the current (2006) tax is $10.30 per pound. The floor stocks tax is $0.45 per pound (2006). For more specifics on the current tax level, see IRS Form 6627, Environmental Taxes.

[7] For example, see Interlaboratory Working Group, Scenarios for a Clean Energy Future, ORNL/CON-476 (November 2000).

[8] For example, see CERA Advisory Service, Design Issues for Market-based Greenhouse Gas Reduction Strategies; Special Report (February 2006), p. 59; Congressional Budget Office, Evaluating the Role of Prices and R&D in Reducing Carbon Dioxide Emissions (September 2006).

[9] Richard D. Morgenstern, Climate Policy Instruments: The Case for the Safety Valve (Council on Foreign Relations, September 20-2 1, 2004), p. 9.

[10] For example, see Dallas Burtraw, David A. Evans, Alan Krupnick, Karen Palmer, and Russell Toth, Economics of Pollution Trading for SO_2 and NO_x (Resources for the Future, March 2005); David Harrison, Jr., Ex Post

Evaluation of the RECLAIM Emissions Trading Program for the Los Angeles Air Basin (Organization for Economic Co-operation and Development, January 21-22, 2003); and Andrew Aulisi, Alexander E. Farrell, Jonathan Pershing, and Stacy VanDeveer, Greenhouse Gas Emissions Trading in U.S. States: Observations and Lessons from the OTC NOx Budget Program (World Resources Institute, 2005).

[11] John P. Weyant, An Introduction to the Economics of Climate Change Policy (Pew Center on Global Climate Change, July 2000).

[12] See CRS Report 98-738, Global Climate Change: Three Policy Perspectives, by Larry Parker and John Blodgett.

[13] Vicki Arroyo and Neil Strachan, Addressing the Costs of Climate Change Mitigation, presented at the Aspen Workshop: A Climate Policy Framework: Balancing Policy and Politics.

[14] Carbon dioxide (CO_2), methane (CH_4), nitrous oxide (N_2O), sulfur hexafluoride (SF_6), hydrofluorocarbons (HFC), and perfluorocarbons (PFC).

[15] For more on sequestration approaches, see CRS Report RL33801, Direct Carbon Sequestering: Capturing and Storing CO_2, by Peter Folger.

[16] The Regional Greenhouse Gas Initiative (RGGI) is an initiative of currently ten northeastern states to reduce GHG emissions. A signed 2005 memorandum of agreement (MOU) requires the parties to stabilize and then reduce CO_2 emissions from powerplants, implemented through an allowance-based cap-and-trade program. If the allowance price rises above $7, offsets from outside the region may be used for compliance purposes at a 1:1 ratio, with the generator able to cover up to 5% of its emissions. (If below $7, such offsets are discounted 50% and the compliance limit is 3.3% of a generator's emissions. If the allowance price exceeds $10, offsets from international projects could be used to cover up to 20% of a generator's emissions.) For more information, see [http://www.rggi.org].

[17] For a review of these estimates, see CRS Report RL30285, Global Climate Change: Lowering Cost Estimates through Emissions Trading — Some Dynamics and Pitfalls, by Larry Parker (available from the author).

[18] For example, see John Reilly, Marcus Sarofim, Sergey Paltsey, and Ronald G. Prinn, The Role of Non-CO_2 Greenhouse Gases in Climate Policy: Analysis Using the MIT IGSM, MIT Joint Program on the Science and Policy of Global Change, Report No. 114 (August 2004); MIT Joint Program on the Science and Policy of Global Change, Multi-gas Strategies and the Cost of Kyoto, Climate Policy Note 3 (April 2000); Vincent Gitz, Jean-Charles Hourcade, and Philippe Ciais, "The Timing of Biological

Carbon Sequestration and Carbon Abatement in the Energy Sector Under Optimal Strategies Against Climate Risks," 27 *The Energy Journal* 3 (2006), pp. 113-133.

[19] For more on the ETS, see CRS Report RL33581, Climate Change: The European Union's Emissions Trading System (EU-ETS), by Larry Parker.

[20] The National Commission on Energy Policy, Ending the Energy Stalemate: A Bipartisan Strategy to Meet America's Energy Challenges (December 2005), p. 21.

[21] Robert Repetto and Duncan Austin, The Costs of Climate Protection: A Guide for the Perplexed (World Resources Institute, 1997), p. 21.

[22] CERA Advisory Service, Design Issues for Market-based Greenhouse Gas Reduction Strategies: Special Report (February 2006), pp. 54-55.

[23] CERA Advisory Service, Design Issues for Market-based Greenhouse Gas Reduction Strategies: Special Report (February 2006), pp. 54-55; Robert Repetto and Duncan Austin, The Costs of Climate Protection: A Guide for the Perplexed (World Resources Institute, 1997) p. 22.

[24] M. Jaccard and W.D. Montgomery, "Costs of Reducing Greenhouse Gas Emissions in the USA and Canada," 24 *Energy Policy* 10/11 (1996), pp. 889-898.

[25] The Clean Air Act (42 U.S.C. 7401-7626), Section 1 10(c)(5)(C).

[26] U.S. Environmental Protection Agency, Letter to California Independent System Operator Corporation (August 12, 2000).

[27] "RECLAIM Poised for Major Changes," Executive Brief (New York: Evolution Markets), pp. 1-2.

[28] South Coast Air Quality Management District, Governing Board Meeting (January 7, 2005).

[29] See Clean Power Group website, [http://www.eea-inc.com/cleanpower/index.html].

[30] 3P.L. 101-559, Title IV, Section 409 (1990).

[31] Dallas Burtraw, David A. Evans, Alan Krupnick, Karen Palmer, and Russell Toth, Economics Pollution Trading for SO2 and NOx, RFF Discussion Paper 05-05 (March 2005), p. 45.

[32] Henry D. Jacoby and A. Denny Ellerman, The Safety Valve and Climate Policy, MIT Joint Program on the Science and Policy of Global Change, Report No. 83 (July 2002), p. 9.

[33] Henry D. Jacoby and A. Denny Ellerman, The Safety Valve and Climate Policy, MIT Joint Program on the Science and Policy of Global Change, Report No. 83 (July 2002), p. 1.

[34] Richard D. Morgenstern, Climate Policy Instruments: The Case for the Safety Valve, Council on Foreign Relations (September 2004), p. 10. This option is discussed further below.
[35] For more information, see CRS Report RL3 3812, Climate Change: Action by States To Address Greenhouse Gas Emissions, by Jonathan L. Ramseur.
[36] Used in the SO2 trading program, credits are allocated gratis to entities in rough proportion to their historic emissions.
[37] Also called an output-based allocation, credits are allocated gratis to entities in proportion to their relative share of total electricity generation in a recent year.
[38] Dallas Burtraw, Karen Palmer, Ranjit Bharvirkar, and Anthony Paul, The Effect of Allowance Allocation on the Cost of Carbon Emission Trading, RFF Discussion Paper 01-30 (August 2001).
[39] Burtraw et al., Allowance Allocation, p. 30.
[40] Karsten Neuhoff, Auctions for CO2 Allowances — A Straw Man Proposal, University of Cambridge Electricity Policy Research Group (May 2007), pp. 3-6.
[41] The National Commission on Energy Policy, Ending the Energy Stalemate: A Bipartisan Strategy to Meet America's Energy Challenges (December 2005).
[42] Jacoby and Ellerman, The Safety Valve and Climate Policy, p. 1.
[43] Richard D. Morgenstern, The Case for the Safety Valve, p. 10.
[44] For a review of U.S. climate change policy, see CRS Report RL30024, Global Climate Change Policy: Cost, Competitiveness, and Comprehensiveness, by Larry Parker and John E. Blodgett.
[45] S.Amdt. 866, passed by voice vote after a motion to table failed 43-54, June 22, 2005.
[46] For background on this point with respect to climate change, see CRS Report RL30285, Global Climate Change: Lowering Cost Estimates through Emissions Trading — Some Dynamics and Pitfalls, by Larry Parker.

Chapter 2

COMMENTS ON DESIGN ELEMENTS OF A MANDATORY MARKET-BASED GREENHOUSE GAS REGULATORY SYSTEM[*]

In light of concerns that rising concentrations of greenhouse gases in the atmosphere may be affecting the Earth's climate, several Members of Congress and public interest groups have proposed plans to require cuts in the United States' emissions of those gases. Implementing a "cap-and-trade" program is an example of one such proposal. Under such a program, policymakers would establish an overall cap on emissions but allow regulated firms to trade rights to those emissions, called allowances. That trading would provide an incentive for firms that could reduce their emissions most cheaply to sell some of their allowances to firms that faced higher costs to reduce their emissions. Such an approach would help reduce the costs of achieving the emissions cap.

In an effort to lay out some of the key questions and design elements of a national greenhouse gas program, Senators Pete Domenici and Jeff Bingaman in February 2006 issued a white paper, Design Elements of a Mandatory Market-Based Greenhouse Gas Regulatory System. That paper asks four questions:

1. Who is regulated and where?
2. Should the costs of regulation be mitigated for any sector of the economy, through the allocation of allowances without cost? Or, should allowances be distributed by means of an auction? If allowances are allocated, what is the criteria for and method of such allocations?

---

[*] This is an edited, excerpted and augmented edition of a Congressional Budget Office, U.S. Congress publication.

3. Should a U.S. system be designed to eventually allow for trading with other greenhouse gas cap-and-trade systems around the world, such as the Canadian Large Final Emitter system or the European Union emissions trading system?
4. If a key element of the proposed U.S. system is to 'encourage comparable action by other nations that are major trading partners and key contributors to global emissions,' should the design concepts of the National Commission on Energy Policy plan (i.e., to take some actions and then to make further steps contingent on the review of what these other nations do) be part of a mandatory market-based program? If so, how?

Further, the white paper solicits comments on any additional topic related to the design of a mandatory market-based program. The Congressional Budget Office (CBO) has issued several papers that address issues raised in the white paper. [1]

In summary, if policymakers decide to limit emissions of carbon dioxide, the primary greenhouse gas, through a cap-and-trade program, they face a choice about where in the production process to implement the regulation. An "upstream" cap would offer two significant advantages and one potential disadvantage over a "downstream" cap:

- An upstream cap would create economywide incentives for households and businesses to reduce their consumption of carbon-intensive goods and services. As a result, it would reduce emissions at a lower cost than if the cap (and resulting incentives for reduction) had been restricted to one downstream sector, such as the electricity sector.
- The costs and complexity of implementing an upstream cap, which would require regulating a limited number of suppliers of fossil fuels, would be significantly less than that of a comprehensive downstream system, which could potentially entail regulating millions of emitters.
- An upstream cap may not provide an incentive to adopt post-combustion technologies that facilitate the capture and sequestration of carbon emissions. Such an incentive could be created by a downstream system that determined allowance requirements on the basis of monitored emissions. An upstream system could provide incentives for sequestration if firms were allowed to meet their allowance requirements by paying for downstream sequestration.

Capping greenhouse gas emissions would impose costs throughout the economy: entities would pay for those costs in the form of higher prices, reduced profits, and lower wages. At the same time, the pool of allowances would have substantial value to those who hold them. Policymakers would need to decide whether to sell the allowances to regulated firms, to give them away, or to implement a combination of the two.

Selling allowances rather than giving them away would not increase the overall economic costs of the cap-and-trade program but would provide an opportunity to use the allowance revenue to reduce other economic distortions. For example, policymakers could use the new source of revenue to reduce existing taxes that tend to slow economic growth (that is, taxes on productive inputs such as capital and labor); to decrease the federal debt; or to fund other government objectives (which otherwise would rely on taxes on productive inputs). As a result, the level of economic activity could be higher if policymakers sold some of the allowances than if they allocated them all at no cost.

Alternatively, policymakers could give some allowances (at no cost) to select firms or individuals to offset the costs that they would incur under the new regulations. Decisions about compensation are complicated by several factors:

- Determining who bears the costs of the cap is difficult. Regardless of whether allowances are sold or given away, the costs of the cap are distributed throughout the economy based on underlying supply and demand conditions.
- Decisions about allocating allowances can increase the overall costs of achieving the cap if they are linked to decisions that influence current emissions. Basing decisions about allowance allocations on historic amounts of production, consumption, or emissions would avoid that problem.
- The costs of the cap would extend beyond firms and consumers to the federal government. Provided that policymakers wanted the government to at least break even under the cap, they would need to reserve a share of the allowances to offset the government's program-induced costs.
- Workers in carbon-intensive industries, such as coal, cement, or aluminum, would be adversely affected if the cap reduced production of those goods. Allocating allowances (at no cost) to firms in affected industries would be likely to benefit the firms' shareholders but not the firms' workers.

Finally, the inclusion of a safety valve, or limit on the maximum price, in the capand-trade program could help keep the economic costs of the program in line with the expected benefits of reducing emissions.

The remainder of this paper is in the format requested by the authors of the white paper in their call for comments. Each of the four questions has several clarifying questions. The three relevant issues that CBO addresses and the related clarifying questions posed in the white paper are all stated at the top of the page in italics, and CBO's responses follow. The paper concludes with CBO's response to the solicitation for comments on additional design topics.

Issue 1: Who is Regulated and Where?

Specific clarifying questions raised in the white paper:

Is the objective of building a fair, simple, and rational greenhouse gas program best served by an economy-wide approach, or by limiting the program to a few sectors of the economy?

What is the most effective place in the chain of activities to regulate greenhouse gas emissions, both from the perspective of administrative simplicity and program effectiveness?

Deciding where in the production process it would be most effective to place the cap would depend on the particular greenhouse gas in question. The following discussion applies to carbon dioxide, the primary greenhouse gas. [2] An "upstream cap" would limit the amount of fossil fuels introduced into the economy; in contrast, a "downstream cap" would place the cap closer to the point where those fuels are combusted and emissions are released. As discussed below, an upstream cap would be expected to be more cost-effective—that is, it would be more likely to achieve any given amount of emission reductions at a lower cost than a downstream cap. [3]

The advantages offered by an upstream cap are twofold. First, it would entail regulating a relatively small number of entities. Second, it would create economywide incentives to reduce the amount of fossil fuel consumed. Thus, it would provide an incentive to cut carbon emissions where they can be reduced most cheaply. (Providing incentives to reduce fossil fuels is equivalent to providing incentives to reduce carbon emissions with one exception—it does not provide an incentive to adopt post-combustion technologies that facilitate the capture of carbon emissions for sequestration. That is discussed in more detail below.)

The economywide incentives for reducing carbon emissions under an upstream design stem from the price increases that would result from limiting the production of fossil fuels. Carbon is a component of fossil fuels. It enters the economy when those fuels are imported or produced domestically and is emitted when they are burned. Under an upstream program, the producers and importers of fossil fuels would be required to hold allowances based on the carbon content of their fuel—that is, the carbon emitted when the fuel is combusted. [4] An upstream cap on carbon emissions would limit production of carbon-based fossil fuels and would cause the price of those fuels to rise—with price increases reflecting each fuel's allowance requirements and, hence, its carbon content.

The increases in fossil fuel prices that would result from the upstream cap would raise firms' and households' costs, encouraging them to decrease their consumption of fossil fuels and energy-intensive goods and services. (For example, households might drive less, and utilities might replace coal with lower-carbon-emitting fuels, such as natural gas or renewable sources of energy.) As a result, households and businesses throughout the economy would have an incentive to reduce all forms of carbon consumption and thus carbon emissions. That equal incentive—throughout the entire economy—would help limit the costs associated with achieving any given level of emission reductions. Further, the higher fossil fuel prices that would result from the cap would provide an incentive for firms to conduct research that could lead to innovations that would reduce fossil fuel use—for example, improvements in energy efficiency and renewable energy sources. Because the price increases would be economywide under an upstream cap-and-trade program, the incentives for innovation would be economywide as well, covering transportation, electricity generation, and industrial processes. As such, an upstream cap-and-trade program could encourage research and development on a wide range of carbon-reducing technologies.

An attempt to achieve economywide incentives for reducing carbon emissions under a downstream cap-and-trade program would probably entail much higher implementation costs. The costs of implementing an upstream program are held down because there is a limited number of producers and importers of fossil fuels and because their allowance requirements could be determined on the basis of information about the amount and type of fuel that they sold in the United States. [5] In contrast, a comprehensive downstream system could entail regulating many more entities. The further downstream the allowance requirement is placed, the larger the number of entities that would need to be regulated. Ultimately, carbon is emitted by roughly 380,000 industrial establishments, millions of commercial buildings, and hundreds of millions of homes and automobiles. [6]

Although an economywide approach to reducing emissions would probably be more cost-effective, the administrative costs of implementing a downstream capand-trade program could be reduced if the cap covered only a limited number of sectors. Roughly 40 percent of carbon dioxide emissions stem from the combustion of coal and natural gas to generate electricity; 32 percent result from the combustion of transportation fuels, such as gasoline and diesel; and the remaining 28 percent stem from the combustion of coal, oil, or natural gas directly by the residential, commercial, or industrial sectors. [7]

Some legislative proposals would have limited a carbon cap to the electricity sector. Limiting the cap to the electricity sector would greatly reduce implementation costs relative to a comprehensive downstream cap; however, it would have several disadvantages relative to an upstream cap. First, a downstream system that was limited to electricity generators would confine incentives for cutting carbon emissions—and for innovation—to that sector, even if potentially lower-cost reductions could have been obtained from sources outside that sector. For example, such a cap would not encourage emission reductions that stem from transportation or from fossil fuel uses in industrial and commercial sectors not associated with their purchase of electricity from covered generators. Second, a downstream cap would offer less certainty than an upstream cap that any desired reduction in U.S. emissions would be achieved. Because the cap would restrict emissions in only one sector, emissions in other sectors could continue to grow. Further, to the extent that electricity generation could shift among establishments to avoid the cap, the system could create leakage. For example, if the program was designed to cover emitters above a certain size (in order to limit the number of regulated entities and to hold down the administrative costs of the program), more electricity generation could shift to facilities that were smaller than the cutoff size.

Although moving the allowance requirement downstream is likely to either increase the costs of implementing the program (if a downstream program was comprehensive) or decrease the cost-effectiveness of the emission reductions that were achieved (if the downstream program was limited to specific sectors), it could offer one advantage relative to an upstream design: it could provide incentives for the use of post-combustion technologies designed to capture carbon emissions for sequestration (that is, long-term storage). That incentive would be achieved if the downstream system regulated actual emissions from sources rather than approximating their emissions on the basis of the fuels that they consume. For example, a cap on emissions from the electricity sector would provide generators with an incentive to install technologies that would scrub emissions from their smokestacks. Those emissions could then be sequestered. (For example, researchers are exploring the feasibility of sequestering carbon emissions in abandoned oil wells and in the ocean.) Alternatively, upstream cap-and-trade programs could be designed to provide incentives for such carbon capture and sequestration if upstream firms were allowed to meet some fraction of their allowance requirement by paying for the capture and sequestration of carbon. (As discussed under clarifying question 2d, those provisions could allow for biological sequestration as well.)

Issue 2: Should the costs of regulation be mitigated for any sector of the economy, through the allocation of allowances without cost? Or, should allowances be distributed by means of an auction? If allowances are allocated, what is the criteria for and method of such allocations?

A general discussion of each of these three questions is provided here. The observations made here apply generally to all of the clarifying questions (about allocations for specific purposes) that follow. Only details that pertain to particular clarifying questions are added under the clarifying questions that begin with 2a: Technology R&D and Incentives.

Should the costs of regulation be mitigated for any sector of the economy, through the allocation of allowances without cost?

Restricting carbon emissions through a cap-and-trade program would probably be costly. As a result, discussions about such a program often include a consideration of whether entities that would bear a particularly large share of that cost would be compensated. (When examining the pros and cons of providing compensation, CBO assumes that decisions about the stringency of the cap would be made independently of decisions about compensation—that is, providing compensation would not be linked to a more stringent cap.) One method of compensating adversely affected entities would be to give them allowances at no cost. Unfortunately, identifying which entities are likely to bear the costs of the cap is difficult. Households, firms, nonprofit organizations, and government agencies all contribute to emissions of carbon dioxide and other greenhouse gases, and all would bear some share of the costs associated with restricting emissions.

Knowing where the cap is placed—that is, which firms would actually be required to hold allowances—provides little insight into who would actually bear the costs of the cap. That is because the costs of the cap do not stick to the point where it is placed; rather, the actual costs of restricting emissions are distributed throughout the entire economy. The extent to which the costs of the cap would be passed forward on to the ultimate consumers of goods and services (such as households and businesses that consume gasoline and electricity) or backward on to fossil fuel suppliers (such as coal producers and oil importers) would depend on the underlying supply and demand conditions for those products. In sum, decisions about which entities might receive compensation are complicated by the difficult task of determining where the actual costs of the cap would land.

Decisions about compensation are unrelated to the decision about where the cap is actually placed because the distribution of the costs of the cap does not depend on the latter decision.

How would allocating allowances at no cost provide compensation? Because a cap-and-trade program would limit the quantity of carbon emissions that are allowed, the right to emit carbon (that is, the allowances) would be valuable. Depending on how stringent the cap is (and thus how valuable the allowances are), that value could be quite large. [8] Policymakers could give entities (for example, households, electric utilities, or coal producers) a share of the allowances to compensate them for the higher costs that they would incur as a result of the cap. Those entities could sell the allowances (to the firms that would be required to hold them) or use them to meet their own allowance requirement (if they are regulated).

Although providing allowances at no cost could compensate some entities, the value of the allowances is going to fall short of the costs that all affected entities combined incur as a result of the cap. [9] As such, policymakers would not be able to offset all firms, households, workers, nonprofits, and government agencies for the costs that they would incur. A decision to provide more compensation to some set of entities would inevitably reduce the compensation that could be offered to others.

Compensation could offset the initial costs of the cap for some entities, but it would not alter the initial distribution of the costs of the cap throughout the economy—that is, it would not alter the ultimate price changes that would result from the cap. [10] For example, providing allowances at no cost to coal producers would not lead to lower coal prices. Thus, compensating coal producers would not protect coal-fired electricity producers, or their customers, from the higher prices that they would be likely to face as a result of the cap. Because compensating entities that are required to hold allowances would probably not affect the price increases that would result from the cap, decisions about compensation would not alter the effect that the policy might have on the competitiveness of U.S. goods.

Difficulties in identifying who actually bears the costs of the cap mean that the government could unintentionally undercompensate or overcompensate various entities. For example, the distributional effects of a cap-and-trade program on electricity producers and consumers would depend on a variety of factors, including the degree of competition in the electricity market, the method of allowance allocation (discussed below), and the mix of generation assets (for example, coal, natural gas, nuclear, and hydro). Effects on an individual utility will differ from effects on the electricity sector as a whole depending on whether it sells power in a regulated or competitive power region, its particular mix of generation assets, and whether the individual entity was in existence when the policy went into effect. [11] Some utilities would be better off, and some would be worse off. Because it is difficult to determine the costs that any given utility would actually bear as a result of the cap, it is also difficult to determine the degree of compensation required to offset those costs, and hence, overcompensation is a possibility.

When examining who actually bears the costs of the cap and considering the possibility of providing compensation, policymakers could consider the costs that the cap would impose on the government. If policymakers wanted the government to at least break even as a result of the cap-and-trade program, they would need to reserve a share of the allowances to offset the costs that the cap itself could impose on the government. [12] Those potential costs stem from several sources. First, governments are consumers of energy and energy-related services. [13] As such, they would bear a share of the costs of a cap-and-trade program that led to higher energy prices. Second, to the extent that the policy reduced economic activity (for example, the gross domestic product), government tax receipts would be reduced. [14] Third, government expenditures for transfer payments linked to price indexes (such as Social Security payments) would increase as a result of policy-induced price increases.

Should allowances by distributed by means of an auction?

As an alternative to distributing allowances without cost, policymakers could sell some, or all, of the allowances. Doing so would provide policymakers with a new source of revenue that could be used to reduce reliance on existing sources of revenue that tend to reduce economic activity. [15]

Most sources of government revenue create unwanted effects—that is, they discourage productive activity. For example, taxes on labor, capital, or income (a combination of the returns to labor and capital) tend to reduce incentives to work and to invest. [16] Selling the allowances would provide a new source of revenue that could be used for a variety of purposes, including reducing existing taxes on productive inputs (such as capital and labor), decreasing the federal debt, or funding other government objectives (which otherwise would rely on taxes on productive inputs).

Thus, although selling allowances (as opposed to giving them away) would not have a direct influence on the costs of the cap, it would create an opportunity for policymakers to use the allowance value to reduce costs associated with unrelated spending or taxation programs. [17] The ultimate economic impact of selling allowances would depend on how policymakers used the allowance revenue. If policymakers gave the revenue back to regulated entities in a lump-sum fashion (not related to their use of capital or labor or their current level of production), the overall economic effect would be equivalent to a program in which they gave allowances to producers at no cost.

Even if the initial allowances (corresponding to the amount of emissions allowed under the cap) were allocated at no cost, the inclusion of a "safety valve" in a capand-trade program could result in the government selling additional allowances. The safety valve would establish an upper limit on the price of allowances. If the price of allowances rose to the safety-valve price, the government would sell as many allowances as was necessary to maintain that price. The amount of allowances sold under such a program would depend on the difference between the stringency of the cap and the safety-valve price. A stringent cap with a low safety-valve price could cause regulated entities to buy a substantial number of allowances.

If allowances are allocated, what is the criteria for and method of such allocation?

Two alternative methods of allocating allowances to firms are "output-based allocations," which link allocations to current production decisions, and "grandfathering," which bases allocations on historic emissions or production decisions. In general, analysts find that grandfathering would result in lower costs than output-based allocations. That is because output-based allocations distort production decisions in ways that increase the costs of obtaining a given level of emission reductions. This issue is discussed in more detail in clarifying question 2f.

Clarifying Questions 2a:

- Technology R&D and Incentives
- What level of resources should be devoted to stimulating technology innovation and early deployment?

- What portion, if any, of the revenues from permits or the auction of allowances should be reserved for technology development? If some portion is reserved for this purpose, should that set-aside flow to the federal government with funds spent through the traditional appropriation process? Or should the funds be allocated directly to a non-profit research consortium, chartered by the federal government, which would then administer technology development and deployment projects? Or should there be some combination of these two options?
- What criteria should be used to determine how such funds are spent and which projects are chosen?
- What other mechanisms should be used to promote technology deployment? Options include tax credits, cost-sharing for demonstration projects, assistance to state energy programs, etc.

Technological advances could play an important role in reducing greenhouse gases at an affordable cost. A cap-and-trade program would provide incentives for firms to invest in developing new technologies; however, firms may not be able to reap the full benefits from those investments. As a result, firms' investments may fall short of the amount that would occur if all of the resulting benefits were taken into account. That shortfall may provide a justification for federal subsidies for R&D.

A cap-and-trade program would place an implicit price on carbon emissions, raising the costs of producing and consuming goods that generate those emissions. The higher prices created by those caps make it profitable for firms to develop technologies that could reduce the costs of cutting carbon emissions. [18] Those innovations could include improvements in energy efficiency or improvements in alternative energy technologies, such as solar, wind, or hydrogen. (Incentives for sequestering carbon would be created only if firms were allowed to meet their allowance requirement by engaging in sequestration activities.) Thus, a cap-and-trade program is appropriately viewed as stimulating private R&D on carbon-reducing technologies.

The magnitude of the incentives for R&D would depend on the stringency of the cap over time. A cap that was implemented for a short period of time would create less incentive for investment in the development of new technologies than one that was expected to persist well into the future. In addition, the more stringent the initial cap was, or future caps were expected to be, the greater would be the incentives for R&D. Because decisions about investing in developing new technologies depend primarily on the future market for those technologies (when the R&D investments would bear fruit), expectations about future caps are of primary importance.

In general, research and development for all technologies (including carbon-reducing technologies) create "spillover benefits"— benefits that society as a whole would receive as a result of a firm's R&D effort but that the firm would be unlikely to capture in the form of higher profits. For example, the development of a new technology may result in general knowledge that is useful in many ways but is not directly covered by a patent. Similarly, one innovation may inspire subsequent innovations that are not tied closely enough to the initial innovation that they are covered by the patent. As a result of those spillover benefits, the profit motive may provide firms with too little incentive to invest in R&D. Existing general tax credits for R&D expenses and current funding of low-carbon energy sources, such as solar, nuclear, and wind, provide some additional incentives to at least partially account for those spillover benefits.

Supplementing private R&D efforts with federal funds would involve both costs and benefits. It would be efficient to the extent that the amount of private R&D stimulated by a cap-and-trade program would fall short of the amount that would occur if all benefits were taken into account. The ultimate efficiency of federal funding would, in turn, depend on the design of the federal funding initiatives (such as investment tax credits, targeted funding of specific technologies, or the offering of federal prizes for technological breakthroughs). The potential costs of federal R&D efforts include the cost of raising funds, the cost of efforts that are ultimately unsuccessful, and the extent to which federally funded R&D on carbon-reducing investments would crowd out other forms of R&D. Thus, it is possible to invest either too much or too little in federal R&D.

The existence of spillover benefits creates an economic rationale for subsidizing R&D on carbon-reducing technologies. However, there is no economic reason to link decisions about funding R&D to the revenues that might be generated by selling allowances under a cap-and-trade program. As described in the general discussion above regarding allocation decisions, the revenue from selling allowances could be used for a variety of different purposes that would have different overall effects on the economy. Likewise, decisions about funding R&D for carbon-reducing technologies could be considered on the basis of their own merit.

Clarifying Questions 2b:

Adaptation Assistance

- What portion of the overall allowance pool should be dedicated to adaptation research or adaptation-related activities?
- How should these allowances or funds be administered?
- What is the appropriate division between federal vs. regional, state, and local initiatives?

In light of the potential for future changes in climate, even if emissions were severely restricted, adaptation could play an important role in any effective climate strategy. [19] The appropriate funding for adaptation could be considered on its own merits—there is no economic reason to link it to the existence of a cap-and-trade program, to link it to the value of allowances created by a cap-and-trade program, or to fund it out of allowance revenues.

Clarifying Questions 2c:

Consumer Protections

- What portion of the allowance pool should be reserved to assist consumers?
- Should funds from the sale of permits or allowances be targeted primarily to low-income consumers, or should they be more widely distributed to benefit all consumers?

A cap-and-trade program is likely to result in higher prices for energy and energy- intensive goods and services as the costs of the carbon restriction are passed on to the ultimate consumers of the products whose consumption results in carbon emissions. Those higher prices play an important role in inducing the behavioral changes that would ultimately reduce emissions, such as using more energy- efficient appliances and purchasing more fuel-efficient cars. At the same time, those higher prices will impose financial costs on consumers. The costs that individual consumers would bear would depend on the amount, and the mix, of goods that they buy. In general, higher-income households would bear more costs (measured in dollar amounts) simply because they consume more goods. Measured as a share of household income, however, the higher prices would impose a larger burden on lower-income households because lower-income households tend to consume a larger proportion of their income. [20]

Clarifying Questions 2d:

What portion of the allowance pool should be reserved for the early reduction credit program and the offset pilot program?
Are other set-aside programs needed? Early Reduction Credits:

A program for reporting voluntary reductions in greenhouse gas emissions has been in effect since 1994. Over 2.5 billion metric tons of emission reductions (measured in carbon dioxide equivalent tons) have been reported under that program in the 1994-2004 time period. [21] The extent to which firms would benefit from those early reductions would depend, in part, on whether allowances were sold. If allowances were sold, early reducers would receive some benefit from their actions because those reductions would decrease the number of allowances that they would need to purchase once the cap was in effect.

If allowances were distributed without cost, then the extent to which firms would benefit from early reductions would depend on whether policymakers allowed them to receive credits for their early reductions (or for a fraction of them). Issuing credits for early reductions would shift costs from companies that engaged in early reductions to ones that did not (provided that the overall cap was unaffected by the amount of early reductions made). Free allowances to early reducers would decrease the number of allowances that could be distributed to other firms. As a result, firms that did not make early reductions would bear a larger share of the costs of meeting the limit on emissions once the cap was in place.

The shift in the cost burden away from firms that received early-reduction credits could be problematic if those credits were earned for reductions that the firms would have found it profitable to make anyway, regardless of regulatory incentives. In that case, companies would receive credit for such reductions, even though they would not have decreased emissions relative to the level that would have occurred without an early-reduction program. [22]

Offsets:

Policymakers would need to decide whether to build incentives for sequestration into a cap-and-trade program. A trading program that calculated allowance requirements on the basis of the carbon content of the fossil fuel used, produced, or sold by a firm would not provide incentives for any form of sequestration. Thus, an upstream program would not provide such incentives. A downstream program could provide some incentives for sequestration, but only if allowance requirements were based on actual emissions. For example, consider a downstream trading program that required electricity generators to obtain allowances. That trading program could provide incentives for installing scrubbers that would capture and sequester carbon emissions, but only if generators' allowance requirements were based on their actual emissions. No such incentive would exist if allowance requirements were estimated on the basis of generators' fuel consumption.

Although a downstream program in which allowance requirements were based on actual emissions could provide incentives for some forms of sequestration, it would not provide incentives for other forms. For example, it would not provide any incentive for firms to offset their emissions with biological sequestration (such as growing trees). Policymakers could build in incentives for biological sequestration by allowing firms to meet some fraction of their allowance requirement by funding such initiatives. Although such sequestration projects could offer low-cost carbon reductions, they could also add considerably to the program's complexity and implementation costs because measuring, monitoring, and enforcing sequestration projects would be difficult.

Clarifying Questions 2e:

Special considerations for fossil-fuel producers?

- Would some upstream fossil fuel producers be unable to pass the cost of purchasing permits or allowances through in fuel prices if they are the regulated entity?
- Is there a sufficient policy rationale for addressing these costs to justify the complexity of setting up and administering an allocation system for these entities?
- What other options exist to address the inability of fossil fuel producers to pass through these costs?

Carbon emissions result from the combustion of fossil fuels, with some fuels leading to greater carbon emissions than others. For example, the amount of carbon released per million British thermal units (Btus) of coal is 1.8 times the amount released per million Btus of natural gas. Differences in the carbon content among fuels mean that some fossil fuel producers and suppliers could be better off as a result of the cap-and-trade program whereas others could be worse off. For example, natural gas producers could be better off if the policy caused electricity generators to switch from carbon-intensive coal to relatively less carbon-intensive natural gas. As a result, the natural gas industry could potentially experience increased profits and higher wages under an initial adjustment period. In contrast, the policy would probably decrease the demand for coal. Therefore, that industry could experience lower profits, decreased wages, and lost jobs, particularly as the industry adjusts to lower output levels. Assuming that allowances were granted on the basis of historic factors (such as a firm's previous production), the granting of allowances would not affect firms' future marginal costs or future production decisions. As a result, compensation provided to firms would be likely to benefit shareholders (it would be equivalent to a windfall gain) but would not be likely to reduce the costs borne by workers because it would not offset the decrease in production that the cap would induce.

The costs that fossil fuel producers would bear as a result of the cap would depend on underlying supply and demand conditions, not on whether they were the regulated entity—that is, required to hold allowances (this point is explained in more detail in the general observations following question 2). As such, the decision about whether to compensate fossil fuel producers (shareholders) or workers need not depend on whether they were the regulated entity.

Clarifying Questions 2f:

Allocations for downstream electric generators?

- Should electricity generators be included in the allocation if they are not regulated? (Clarification: We mean to ask if an electric generator should be included in the allocation if the greenhouse gas regulation occurs at a point of regulation that is upstream or downstream from the generator, but not the generator itself.)
- What portion of the total allocation should be granted to the electric power sector? Should it be based on the industry's share of greenhouse gas emissions or some other factor?
- Should generators in competitive and cost-of-service markets be treated differently under an allocation scheme?
- How should permits or allowances be distributed within the electric sector? Should it be based on historic emissions? Electricity output? Heat input?

As observed in the general discussion following question 2 above, the costs that entities would bear under a cap-and-trade program generally depend on the underlying conditions of supply and demand, not on whether those entities are required to obtain allowances for their emissions. As a result, decisions about whether to compensate electricity generators need not be linked to decisions about whether the allowance requirement is placed on them or upstream or downstream of them.

However, provided that policymakers decided to place the allowance requirement on electricity generators, there could be a reason why selling allowances to generators in cost-of-service markets would be more efficient than issuing them at no cost. In most cases, regulators include inputs at their "original cost" (actual prices paid for them) when calculating electricity prices. [23] As a result, allowances that generators receive at no cost would not lead to higher electricity prices in cost-of-service markets. (In competitive markets, that would not be the case because firms would reflect the opportunity cost of using the allowance—that is, the forgone revenue from not selling it—in the prices that they charge for electricity.) Failure to pass the opportunity cost of using allowances on to electricity customers, however, would provide consumers in cost-of-service markets with an insufficient incentive to reduce their use of electricity. As a result, allocating allowances at no cost to electricity generators in cost-of-service markets could increase the cost of reducing emissions, and auctioning allowances to generators in those markets would be more efficient.

The cap-and-trade program could impose higher costs on electricity generators, particularly those that burn coal. Those generators, however, are likely to pass much of those costs on to their customers in the form of higher prices (as discussed above, this is particularly likely in electricity markets with a high degree of competition). As a result, matching the industry's share of allowances to its share of greenhouse gas emissions would probably overcompensate generators because much of the cost of the cap would be passed on to electricity consumers.

The costs that an individual utility would bear would depend on whether it sells power in a regulated or competitive power region, its particular mix of generation assets, and whether it was in existence when the policy went into effect. Some utilities would be better off, and some would be worse off. [24] Efforts to match compensation to actual costs would have to take those factors into account.

Basing allocations on current production decisions (called output-based allocations) rather than on historic emissions or production decisions (called grandfathering) could increase the overall costs of meeting an emissions cap. A cap-and-trade program would be most effective at reducing the costs of attaining an emissions cap if the trading program provided equal incentives for businesses and households to engage in all forms of carbon-reducing activities: it should not provide greater incentives for some activities than for others. Provided that electricity is sold in a competitive market, a cap-and-trade program in which allowances (or a share of them) were grandfathered to existing firms would meet that condition, whereas a program in which allowances (or a share of them) were allocated to firms on the basis of their current production would not.

Allocations that were linked to historic emissions or production decisions would not affect electricity producers' future production decisions or future electricity prices. Thus, the costs associated with emitting carbon would be passed on to the firms and households that use electricity, providing them with an incentive to limit their use. In contrast, output-based allocations would link a producer's allowance allocation to its current production decisions. Thus, the costs of producing a unit of electricity would be subsidized by the allowances earned as a result of the additional production. As a result, output-based allocations would tend to encourage more electricity production and lower electricity prices. Lower electricity prices, in turn, would mean that the policy would not give firms and households as much incentive to limit their electricity use. Although output-based allocations could lower the costs that the cap-and-trade program would impose on electricity consumers, it would increase the overall costs of the program. Higher- cost emission reductions in other sectors would need to make up for the increased emissions (relative to grandfathering) in the electricity sector. [25]

Clarifying Questions 2g:

Allocations for energy-intensive industries?

- Is there a sufficient policy rationale to have an allocation to selected energy-intensive industries? What industries should be included in the allocation?
- What portion of the overall allocation framework should be reserved for these industries?
- What are the appropriate metrics for determining allocations across different industries?

A restriction on carbon emissions would lead to higher energy prices and thus would impose costs on energy-intensive industries such as steel, aluminum, chemicals, pulp and paper, and cement. Higher production costs for those industries would tend to decrease their competitiveness, particularly if the prices for their goods were determined in world markets (where higher production costs could not be passed on to consumers in the form of higher prices). As a result of those higher production costs, production levels, profits, and wages in those industries could decline.

Giving allowances to firms in energy-intensive industries could compensate shareholders for the reduction in profits. However, assuming that allowances were granted on the basis of historic factors (such as a firm's previous production), such allowances would not offset any reduction in the competitiveness of those industries because they would not lower the costs of producing the energy-intensive goods. [26] Correspondingly, giving allowances would not offset the costs that workers in those industries might bear as a result of the decrease in production.

Clarifying Questions 2h:

Allocations to other industries/entities?

- What other industries/entities (e.g. agriculture, small businesses, etc.) allowances considered in the allocation pool?
- What should be the basis for their share of the total allocation as well as for the distinction among such industries/entities?

The Congressional Budget Office has not written about this issue in the past and, as a result, has not offered a response to these questions.

Issue 3: Should a U.S. system be designed to eventually allow for trading with other greenhouse gas cap-and-trade systems around the world, such as the Canadian Large Final Emitter system or the European Union emissions trading system?

Clarifying questions raised in white paper:

- Do the potential benefits of leaving the door open to linkage outweigh the potential difficulties?
- If linkage is desirable, what would the process for deciding whether and how to link to systems in other countries?
- What sort of institutions or coordination would be required between linked systems?

Because emissions from anywhere in the world make the same potential contribution to warming, a mitigation program would minimize the costs of meeting any particular goal by placing the same price on emissions everywhere. Thus, if policymakers were to adopt cost-effectiveness as a guiding principle in controlling emissions, they would want to ensure that emission prices would be equalized across countries. One way to accomplish that goal would be to allow for the trading of emission credits or rights across international borders.

Nevertheless, international trading could raise or lower the domestic price of emissions and the overall costs of the domestic program, depending on what set of countries was included in the system and the relative stringency of participating countries' domestic programs. For example, if trading only involved developed countries, each with an emission target that required similarly proportionate reductions in baseline emissions, emissions trading would be likely to raise prices in the United States, benefiting owners of domestic emission credits but hurting fuel users. In contrast, if a trading system included developing countries such as India and China, and those countries had targets consistent with their projected baseline emissions, emissions trading could result in a dramatic decrease in the emission price in the United States. [27]

Further complications would arise if cap-and-trade systems in different countries had dramatically different rules. Significant variations among systems would be likely to significantly increase monitoring and enforcement costs. Even more complications in monitoring and enforcement would arise if a domestic trading system allowed for regulated entities to earn credits by sponsoring emission- reducing projects in countries that did not have any targets at all. Further, countries' ability to ensure that their emission target would be met could be limited if any participating country's system incorporated a safety valve, or limit on the maximum price, and if regulated entities in other countries were allowed access to credits available at the safety-valve price. For example, if the clearing price for emission allowances necessary to meet the cap in the European Union trading program was higher than a safety-valve price included in a U.S. trading program, then European firms could comply by purchasing U.S. allowances. If that was to occur, the emissions cap in the EU program would not be met.

> If there is an additional topic related to the design of a mandatory market based program that you would like to address, please submit comments on this form.

A cap-and-trade program for carbon dioxide emissions would offer a way to set an overall limit on the level of carbon dioxide emissions while relying on economic incentives to determine where and how emission reductions occur. Such a program would probably reduce the costs of meeting an emission- reduction target, but it would not necessarily balance actual costs with the expected benefits achieved by the target. As described below, including a "safety valve" in a cap-and-trade program could help achieve that goal.

A cap-and-trade program with a safety valve would combine an overall cap on total emissions with a ceiling on the allowance price. If the price of allowances rose to the ceiling (or safety-valve) price, the government would sell as many allowances as was necessary to maintain that price. Thus, if the safety valve was triggered, the actual level of emissions would exceed the cap. The cap would be met only if the price of allowances never rose above the safety-valve price.

If policymakers had complete and accurate information on both the costs and benefits of achieving various limits on emissions, the inclusion of a safety valve would not offer any economic advantages. With full information, policymakers could set the cap to the level at which the cost of the last ton of emissions reduced in order to meet the cap was equal to the benefit from that reduction. However, neither the costs nor the benefits are known with certainty. For that reason, the best policymakers can do is to choose the policy instrument that is most likely to reduce the cost of making a "wrong" choice. Choosing a cap that is too stringent would result in excess costs that are not justified by their benefits. The inclusion of a safety valve that limited the price of allowances to the expected benefits of incremental emission reductions would avoid that outcome.

The advantages of including a safety valve in a cap-and-trade program stem mainly from the fact that the cost of limiting a ton of emissions is expected to rise as the cap becomes more stringent, whereas the expected benefit of each ton of carbon dioxide reduced is roughly constant across the range of potential emission reductions in a given year. [28] Because the additional benefit created by each additional ton of carbon that is reduced as the cap is tightened is expected to remain constant (even though it cannot be known with certainty), yet the additional cost is expected to rise by an unknown amount, a safety valve could help prevent excess costs. A safety valve would limit the cost of additional emission reductions to the expected benefit of those emission reductions. [29]

REFERENCE

[1] See Congressional Budget Office, Who Gains and Who Pays Under Carbon-Allowance Trading? The Distributional Effects of Alternative Policy Designs (June 2000); An Evaluation of Cap-andTrade Programs for Reducing U.S. Carbon Emissions (June 2001); The Economics of Climate Change: A Primer (April 2003); The Economic Costs of Reducing Emissions of Greenhouse Gases: A Survey of Economic Models (May 2003); Uncertainty in Analyzing Climate Change: Policy Implications (January 2005); and Limiting Carbon Dioxide Emissions: Prices Versus Caps (March 15, 2005).

[2] In 2004, carbon dioxide from energy combustion accounted for 82.4 percent of greenhouse gas emissions in the United States (measured in carbon dioxide equivalents). See the Energy Information Administration's annual reports on U.S. greenhouse gas emissions at www.eia.doe.gov/oiaf/1605/1605aold.html.

[3] For a more detailed discussion of the pros and cons of an upstream versus a downstream design, see Congressional Budget Office, An Evaluation of Cap-and-Trade Programs for Reducing U.S. Carbon Emissions (June 2001).

[4] To avoid making the cap place U.S. exports of fossil fuels at a disadvantage, fossil fuels that were exported could be exempt from the cap. Regulating importers and exempting exporters would have the effect of restricting the emissions associated with U.S. consumption (not production) of fossil fuels.

[5] The Center for Clean Air Policy estimated that an upstream program would require less than 2,000 entities to hold allowances. See Center for Clean Air Policy, U.S. Carbon Emissions Trading: Description of an Upstream Approach (Washington, D.C.: Center for Clean Air Policy, March 1988), p. 7.

[6] Ibid., p. 5.

[7] 7. See www.eia.doe.gov/emeu/aer/txt/ptb1202.html.

[8] For example, U.S. entities released rough 7 billion metric tons of greenhouse gases (measured in carbon dioxide equivalents) in 2004. Valued at $7 per ton (the safety-valve price used in the National Commission on Energy Policy report Ending the Energy Stalemate: A Bipartisan Strategy for Meeting America's Energy Challenges), the value of those emissions would be $49 billion.

[9] The costs of the policy would include the costs of the allowances themselves (equivalent to the allowance value) and the substitution costs—

that is, the costs that entities would bear from reducing their consumption of fossil fuels.

[10] Some exceptions to this are if allowances are granted as a function of current production or if allowances are given to utilities whose electricity prices are set by regulators. Those exceptions are discussed in more detail in the following sections.

[11] For a discussion of those factors, see Dallas Burtraw and others, "The Effect on Asset Values of the Allocation of Carbon Dioxide Emission Allowances," The Electricity Journal, vol. 15., no. 5 (Washington, D.C.: Resources for the Future, March 2002).

[12] This discussion does not include the costs of actually implementing the cap-and-trade program.

[13] The government is estimated to have consumed roughly 13 percent of carbon consumed in the United States in 1998. See Congressional Budget Office, Who Gains and Who Pays Under Carbon-Allowance Trading? The Distributional Effects ofAlternative Policy Designs (June 2000), p. 11.

[14] In contrast, to the extent that the allowance distribution led to increases in shareholders profits, a fraction of that increase would be received by federal, state, and local governments through collections in taxes on profits. For a discussion of the distributional effects that different allocation decisions would have, see Congressional Budget Office, Who Gains and Who Pays Under Carbon- Allowance Trading? and Terry M. Dinan and Diane Lim Rogers, "Distributional Effects of Carbon Allowance Trading: How Government Decisions Determine Winners and Losers," *National Tax Journal*, vol. 55, no. 2 (June 2002), p. 206.

[15] Higher energy prices created by the cap would tend to slow economic growth as well. However, those price increases would occur regardless of whether the government sold the allowances or gave them away.

[16] Higher prices created by a cap on emissions would reduce real income from working and investing and, thus, the incentive to do so. Such reductions in inputs to production would exacerbate the discouraging effect that existing taxes on labor and capital already have on productive activity. The exacerbation of existing tax distortions—called the tax-interaction effect—is difficult to measure but could be significant. However, the magnitude of the tax-interaction effect is likely to be the same whether allowances are sold or given away.

[17] For a discussion of the distributional implications of alternative allocation schemes, see Congressional Budget Office, An Evaluation of Cap-and-Trade Programs for Reducing U.S. Carbon Emissions.

[18] In the absence of an explicit incentive to reduce carbon emissions, firms' incentives to reduce fossil fuel consumption (and associated carbon emissions) would stem from other market forces, such as the rising price of oil due to underlying conditions in supply and demand. Firms' investments in energy efficiency or alternative energy technologies, however, would fall short of the amount that would occur if they had an incentive to take the benefits of reducing carbon emissions into account.

[19] See Congressional Budget Office, Uncertainty in Analyzing Climate Change: Policy Implications (January 2005), p. 36.

[20] See Congressional Budget Office, Who Gains and Who Pays Under Carbon-Allowance Trading?, p. 21, table 4.

[21] To provide perspective, U.S. emissions of greenhouse gases in 2004 are estimated at approximately 7 billion metric tons.

[22] For a discussion of early-reduction crediting, see Congressional Budget Office, An Evaluation of Cap-and-Trade Programs for Reducing U.S. Carbon Emissions, pp. 14-15.

[23] See Dallas Burtraw and others, "The Effect on Asset Values of the Allocation of Carbon Dioxide Emission Allowances," pp. 51-62.

[24] For a discussion of those factors, see Burtraw and others, "The Effect on Asset Values of the Allocation of Carbon Dioxide Emission Allowances."

[25] For a more detailed discussion of grandfathering versus output-based allocations, see Congressional Budget Office, An Evaluation of Cap-and-Trade Programs for Reducing U.S. Carbon Emissions. Also see Dallas Burtraw and others, The Effects of Allowance Allocation and the Cost of Efficiency of Carbon Emission Trading (Washington, D.C.: Resources for the Future, April 2001).

[26] The results would be different if the number of allowances that firms received was directly linked to their current, or future, production decisions (referred to as "output-based" allocations). In that case, firms would "earn" allowances on the basis of their production decisions—and the decline in production that would result from the cap could be less. As described in the discussion under clarifying question 2f, however, such output-based allocations would be inefficient—that is, they would increase the costs of obtaining any given amount of carbon reductions.

[27] See Congressional Budget Office, The Economic Costs of Reducing Emissions of Greenhouse Gases: A Survey of Economic Models (May 2003), p. 82.

[28] That constancy occurs because climate effects are driven by the total amount of carbon dioxide in the atmosphere, and emissions in any given

year are a small portion of that total. Further, reductions in any given year probably would be considerably less than the total baseline emissions for that year.

[29] Limiting emissions of carbon dioxide with a tax on carbon emissions (set equal to the expected benefit of reducing emissions by one ton) could offer additional economic advantages over a cap-and-trade program with a safety valve. If the costs of reducing emissions were greater than expected, the tax would perform in the same manner as the safety valve. However, if the costs of reducing emissions were less than expected (and thus, the cap was less stringent than might have been justified by actual costs and benefits), the tax could offer additional advantages. The tax could motivate more emission reductions than would have been required by the cap—keeping the cost of emission reductions in line with the benefits that they were expected to create. Available research indicates that a price instrument, such as a tax or safety valve, would offer economic advantages over a cap as long as policymakers did not feel it necessary to make extremely large emission reductions in the near term to avoid passing a threshold level of atmospheric concentration—that is, a point at which incremental increases in emissions would lead to a large increase in the incremental damages caused by those emissions. For a more detailed description of the advantages that a tax and a safety valve offer, along with an illustrative example, see Congressional Budget Office, Limiting Carbon Dioxide Emissions: Prices Verus Caps (March 15, 2005).

In: Designing Greenhouse Gas Reduction… ISBN 978-1- 60741-195-6
Editor: Sonja Enden © 2009 Nova Science Publishers, Inc.

Chapter 3

DESIGN ELEMENTS OF A MANDATORY MARKET-BASED GREENHOUSE GAS REGULATORY SYSTEM

Pete V. Domenici and Jeff Bingaman

PURPOSE

The purpose of this document is to lay out some of the key questions and design elements of a national greenhouse gas program in order to facilitate discussion and the development of consensus around a specific bill. We recognize that there are many ways to structure such a regulatory program and that there are entirely different approaches that might include a carbon tax, technology incentives and voluntary programs, but we have limited our consideration here to "mandatory market-based systems" contemplated by the Sense of the Senate Resolution.

INTRODUCTION

"*Congress finds that—*

(1) greenhouse gases accumulating in the atmosphere are causing average temperatures to rise at a rate outside the range of natural variability and

are posing a substantial risk of rising sea-levels, altered patterns of atmospheric and oceanic circulation, and increased frequency and severity of floods and droughts;

(2) there is a growing scientific consensus that human activity is a substantial cause of greenhouse gas accumulation in the atmosphere; and

(3) mandatory steps will be required to slow or stop the growth of greenhouse gas emissions into the atmosphere.

"It is the sense of the Senate that Congress should enact a comprehensive and effective national program of mandatory, market-based limits and incentives on emissions of greenhouse gases that slow, stop, and reverse the growth of such emissions at a rate and in a manner that—

(1) will not significantly harm the United States economy; and

(2) will encourage comparable action by other nations that are major trading partners and key contributors to global emissions."

The United States Senate adopted this statement of overall national greenhouse gas policy as an amendment to its version of the Energy Policy Act of 2005 on June 22, 2005. We believe it provides the basic framework for the further Senate action on global warming that is necessary to move our energy system into a sustainable and predictable future, to avoid destructive interference with the world climate system, and to maintain long-term U.S. competitiveness and economic prosperity.

The resolution was not included in the EPACT 2005 Conference Report. Nevertheless, we are committed to move the process of Senate action forward in this session of Congress by formulating and advancing proposals that would embody the principles of the Senate Resolution on national climate change policy.

As the Chairman and Ranking Member of the Committee on Energy and Natural Resources, we have already taken action since the June 22 resolution to lay the ground for a national program that would achieve the above objectives. The Committee has held two hearings on scientific and economic considerations relevant to developing such a national program. The committee will also hold a climate change conference this spring to identify the challenges to implementing an effective national program and explore possible solutions to those challenges.

KEY ELEMENTS OF PROPOSALS FOR A NATIONAL PROGRAM

1. Who is Regulated and where?

Two important decisions with respect to who is being regulated in any greenhouse gas program need to be made at the outset. The first decision is whether to build a program that addresses greenhouse gases on an economy-wide basis or whether to build a program that focuses on just the greenhouse gas emissions of one or more industrial sectors. The second decision is where in the "industrial life-cycle" of carbon dioxide and other greenhouse gases should the government choose to implement its regulatory program.

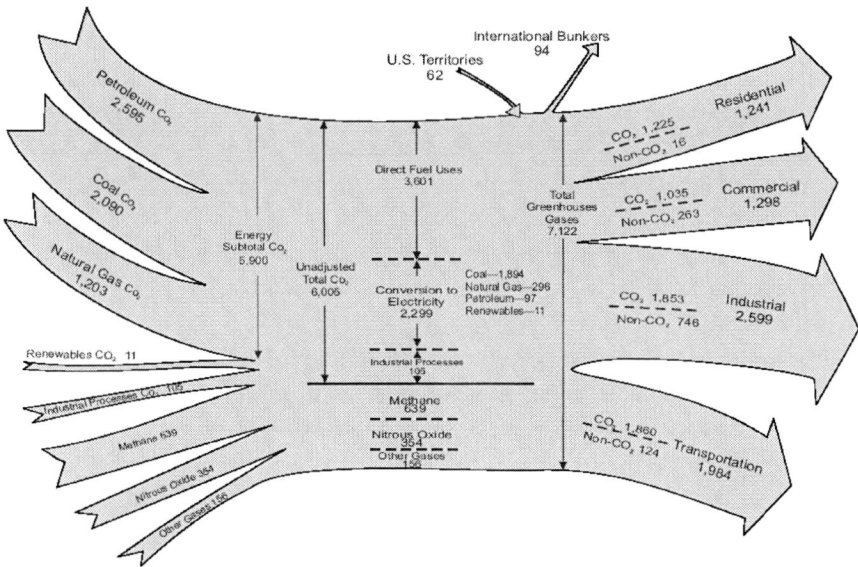

Source: Energy Information Administration, Emissions of Greenhouse Gases in the United States 2004. December 2005

Source: Energy information Administration, Emissions of Green house Gases in the United States 2004. December 2005.

Sectoral versus economy-wide approaches. As the Energy Information Administration chart below shows, there is no single sector of the U.S. economy that makes an overwhelming contribution to overall U.S. greenhouse gas emissions. If a key design feature is fairness, then no one sector should be singled out. An economy-wide approach also allows for ease in seeking the least-cost path to reductions through trading systems. A sector-specific approach might be easier to set up, but leaves open the question of whether it would ever be expanded to include more emitters, or whether further action to address climate change would be unfairly targeted to the sector already under regulation.

"Upstream" versus "downstream" regulatory approaches. Direct emitters of greenhouse gases in the United States number in the hundreds of millions, if one includes sources such as automobiles and residential furnaces. This fact makes it difficult to cover all direct emitters in an economy-wide regulatory program should such an approach be chosen. However, there are various points within the chain of energy production, distribution, and end use where a regulatory requirement to obtain and turn in greenhouse gas permits might be implemented. These include fuel extraction (oil and gas wells and coal mines), processing (oil refineries, natural gas processing facilities, coal blending/cleaning facilities), fuel transportation (pipelines, shippers) or further down the energy chain, such as electric generation, distribution utilities, and large industrial energy users.

In an "upstream" regulatory approach, the point of regulation is placed closer to energy producers and suppliers than to end-use consumers. Specifically, a requirement to acquire permits or allowances for emissions associated with fossil fuel use would apply to coal mining companies, petroleum refiners, and natural gas shippers or pipelines rather than to the "smokestack" entities (e.g., electric utilities, large industrial plants) that, along with households and small businesses, buy and consume these fuels.

There are several arguments that favor this so-called "upstream" approach. First, placing the point-of-regulation relatively high up in the progression from energy production to consumption reduces the number of sources that must be regulated and simplifies program administration. Second, this approach more efficiently captures all sources of emissions and all emissions reduction opportunities throughout the economy. For the same reason, it may stimulate a wider range of emissions reduction responses throughout the economy assuming permit costs are passed on to end users. In addition, an upstream approach may reduce overall administrative costs.

The other basic regulatory option is a "downstream" approach, which targets the regulatory program at the point of emissions. Such an approach likely would be quickly limited to major emitters only, in order to keep the number of covered sources manageable. A program of this type would be best suited to a scenario in which the decision was made to regulate only specific sectors of the economy, large emitters (e.g.,electric generators and energy-intensive industries in the manufacturing sector). Comparable systems include the U.S. acid rain trading program. A chief advantage of this system is that the United States already has significant experience with this approach.

It is hard to see how greenhouse gas emissions from the transportation sector could be addressed in a downstream permitting system. Failing to address transportation would leave out one-third of total U.S. carbon emissions. Most of the ways of addressing transportation that have been proposed are variants of an upstream approach. For example, some have proposed that petroleum refiners be required to hold and turn in permits for the greenhouse gas emission equivalent of all transportation fuels sold. Another option would be to require automobile and truck manufacturers to hold permits covering the carbon emissions from the vehicles they sell, with the emissions estimated based on annual average vehicle miles traveled (VMT) and average vehicle lifetime. Another proposal has been to allow limited trading of increases in fuel economy performance of new cars and trucks, since increased fuel economy requirements could also reduce carbon emissions from the transportation sector. Such a system would have to exclude from the carbon trading program emissions from vehicles currently on the road. This approach would also have to address the problem that improved vehicle efficiency combined with low gasoline prices could lead to an increase in VMT that would offset the carbon dioxide reduction achieved through fuel economy gains.

Key Questions

- Is the objective of building a fair, simple, and rational greenhouse gas program best served by an economy-wide approach, or by limiting the program to a few sectors of the economy?
- What is the most effective place in the chain of activities to regulate greenhouse gas emissions, both from the perspective of administrative simplicity and program effectiveness?

2. Should the Costs of Regulation be Mitigated for any Sector of the Economy, through the Allocation of Allowances without Cost? Or, should Allowances be Distributed by Means of an Auction? If Allowances are Allocated, what is the Criteria for and Method of such Allocation?

Free allowances are not strictly necessary for the operation of a greenhouse gas control program. In fact, free allowances might result in greater cost and complexity for the program. To minimize the costs of a trading program to the U.S. economy as a whole, the government could simply auction all greenhouse gas emission allowances. If all allowances were allocated solely through an auction, then there would be no need to develop and administer allocation rules, and no prospect that such rules might result in unintended competitive advantages, including windfall profits, for certain market participants.

While the allocation of allowances won't reduce the overall cost of the program, it can shift compliance cost among participants. For example, even though in an upstream system, downstream users of carbon or energy, such as electric utilities and energy- intensive manufacturers would not have to buy, sell, or turn in allowances, there might be a rationale for these non-regulated entities to receive allowances that they could then sell to regulated entities. In particular, these entities might face higher energy costs based on greenhouse gas emissions that they are unable to pass through to purchasers of their products. An allowance allocation would thereby allow them to be compensated for these costs.

In addition to the price signal created by the cap and trade program, greater economic incentive could be provided for the development of low-carbon energy technologies by subsidizing those technologies with the revenue from the auction of allowances. Those receipts could also be used to help fund adaptation measures to unavoidable climate change impacts or mitigate costs of the policy to consumers, especially if such additional costs are having a regressive impact on low-income households.

These potential rationales for allowance allocation are explored in more detail below.

a. Technology R&D and Incentives

Virtually all experts agree that significant technology advancements will be needed to adequately and affordably address climate change over the next century. Unfortunately, investments in energy R&D and incentives for the early deployment of advanced technologies currently fall short of what is likely to be needed to tackle this global problem. In its December 2004 report, the National

Commission on Energy Policy (NCEP) called for approximately doubling public sector investments in energy R&D, from $1.7 billion annually (FY 2004) to $3.3 billion per year, and for increasing early deployment incentives from $600 million (FY 2004) to $2 billion annually. Reserving 5 percent to 10 percent of total permit or allowance pool annually for free distribution to technology development and deployment could provide approximately the revenue needed to support the funding increases recommended by NCEP.

Key Questions

- What level of resources should be devoted to stimulating technology innovation and early deployment?
- What portion, if any, of the revenues from permits or the auction of allowances should be reserved for technology development? If some portion is reserved for this purpose, should that set-aside flow to the federal government with funds spent through the traditional appropriation process? Or should the funds be allocated directly to a non-profit research consortium, chartered by the federal government, which would then administer technology development and deployment projects? Or should there be some combination of these two options?
- What criteria should be used to determine how such funds are spent and which projects are chosen?
- What other mechanisms should be used to promote technology deployment? Options include tax credits, cost-sharing for demonstration projects, assistance to state energy programs, etc.

b. Adaptation Assistance

Even very aggressive emission reduction policies undertaken today are unlikely to fully mitigate the impacts of future warming, some of which is almost certain to occur given historic and current levels of global greenhouse gas emissions. Actions to moderate the consequences of climate change — which may include rising sea levels, melting permafrost, altered precipitation patterns, more intense and frequent extreme weather events, and changes to the geographic distribution of important disease vectors — must therefore complement actions aimed at mitigating its causes. Adaptation measures can substantially reduce the potential for damage by improving the ability of human and natural systems to respond to the consequences of climate change. In light of highly varied potential vulnerabilities, a multitude of adaptive policy options exist. Policies like improved flood plain and coastal development zoning could help minimize future property damages, while vigorous agriculture and coastal research programs could better prepare susceptible economic sectors for likely shocks. Adaptation strategies are important components of an integrated approach to the risks posed by climate change, and should be grounded to the extent possible in the best current understanding of likely regional and local effects of climate change. As such, both adaptation research and adaptive policies deserve serious consideration in the near term.

Key Questions

- What portion of the overall allowance pool should be dedicated to adaptation research or adaptation-related activities?
- How should these allowances or funds be administered?
- What is the appropriate division between federal vs. regional, state, and local initiatives?

c. Consumer Protections

While the energy price impacts of any climate change proposal we are currently considering will be quite modest, even small increases in energy prices can disproportionately affect low-income and fixed-income households. Allowances could be used to help offset the cost of higher fossil fuel prices for these consumers. For example, reserving 1 percent of the annual allowance pool for low-income energy assistance could increase annual LIHEAP funding by an average of 20 percent during the first decade of the program. Allowances could also be used to fund additional incentives for consumers to purchase more energy-efficient products that would help consumers reduce their exposure to energy price increases.

Key Questions

- What portion of the overall allocation pool should be reserved to assist consumers?
- Should funds from the sale of permits or allowances be targeted primarily to low- income consumers, or should they be more widely distributed to benefit all consumers?

d. Set-Aside Programs

A portion of permits or allowances could be set aside for an early reduction credit program and an offset pilot program. The early reduction credit program would award permits or allowances to companies or other organizations that reduced emissions prior to the implementation of a mandatory program. These include reductions reported through DOE's 1 605b program, and reductions made through other government-sponsored and private programs identified by the Secretary of Energy.

The offset pilot program would reserve a limited number of permits or allowances from the total allocation to be awarded to entities that achieve greenhouse gas reductions from sources that are not covered under the cap. Example projects might include biological carbon sequestration through forestry and agriculture or capturing methane from landfills. While offset projects can provide a very low-cost and efficient means of achieving reductions, many projects of this type also present significant challenges in terms of measuring, monitoring, and verifying emission reductions. By dedicating a percentage of permits or allowances from within the program's overall emissions budget to offset activities, the nation can undertake a large scale demonstration program aimed at resolving some of these issues while still ensuring that the program achieves its intended environmental goals.

Key Questions

- What portion of the allocation pool should be reserved for the early reduction credit program and the offset pilot program?
- Are other set-aside programs needed?

e. Special Considerations for Fossil-Fuel Producers?

Under an upstream emissions trading program, carbon dioxide emissions would be regulated at the point of fossil fuel production. Regulated firms would be required to submit allowances both for their own emissions and for the carbon dioxide contained in the total amount of fuel each firm sells. The most likely points of regulation in an upstream system are the fossil-fuel producers -- petroleum refiners, natural gas shippers or pipelines, and coal mining companies.

In a downstream regulatory program, fossil-fuel producers are not regulated entities and have no compliance costs.

The compliance costs for fossil fuel producers in an upstream regulatory system, where they are the regulated firms, represent only a small portion of the overall costs of any trading program. Most upstream producers can and will simply pass the value of any allowances they require through to fuel prices, regardless of whether they receive the allowances for free or are required to pay for them. EIA's analysis of the NCEP proposal confirms our expectation that petroleum refiners and natural gas shippers in an upstream system should be able to pass almost all compliance costs through to consumer prices. Coal companies are able to pass a substantial portion of their costs through in prices, although not the full amount.

The ultimate cost borne by fossil fuel producers is therefore far less than the cost of purchasing allowances. Instead, the real cost to producers is a function of three factors: (1) any permit or allowance costs they cannot pass on in fuel prices; (2) any costs associated with their own emissions they cannot pass on in fuel prices; and (3) any loss of revenues due to reductions in demand for fossil fuels. Regarding the first factor, EIA's analysis indicates that permit or allowance prices are likely to be largely passed on to purchasers, with little change in the prices received by producers. Regarding the second factor, EIA estimates that refineries constitute about 20 percent of total manufacturing emissions, with gas and coal production accounting for much less. Finally, to the third point, petroleum consumption falls only slightly as a result of the emissions trading program (1 percent in 2020), while natural gas demand increases slightly. The largest impact is on coal, where demand and coal sales continue to grow, but more slowly than would be projected in the absence of a regulatory program – to a level that is 5 percent lower by 2020 than the projected increased levels, and to a level that is 9 percent lower by 2025. All told, these costs would be offset completely by an allocation of roughly 5 to 10 percent of the total permit or allowance pool to fossil fuel producers.

Key Questions

- Would some upstream fossil fuel producers be unable to pass the cost of purchasing permits or allowances through in fuel prices if they are the regulated entity?
- Is there a sufficient policy rationale for addressing these costs to justify the complexity of setting up and administering an allocation system for these entities?
- What other options exist to address the inability of fossil fuel producers to pass through these costs?

f. Allocations for Downstream Electric Generators?

Although electric generators would not be regulated under an upstream regulatory program, they will face higher production costs as fossil fuel prices rise. A portion, though not all, of these additional fuel costs will be passed through in higher electricity prices. To the extent that generators receive allocations of free allowances, they can sell those allowances and use the revenue to offset higher fuel costs.

Would there be a reasonable policy rationale for such an allocation? There certainly would not appear to be one for existing non-fossil electric generation facilities. These generators do not face higher costs as a result of the program, and as prices for fossil- generated electricity rise in response to the trading program, they will receive more revenue for their generation. Moving forward, however, new non-fossil units could be included in the allocation as one means of overcoming current barriers to new investment in non-emitting generation.

The economic analysis undertaken of the NCEP proposal by the EIA suggests that a 10 percent share of the total allocation would fully offset adverse impacts on electric generators. (Since the electric sector is responsible for 40 percent of national carbon emissions, 10 percent of the total allowance pool equates to 25 percent of the carbon content of fuel consumed by electric generators). The 10 percent figure assumes that the allocation system perfectly targets allowances to the companies that bear nonrecoverable costs. Recognizing that a perfectly targeted allocation is not possible and that some "passed through" costs will revert to fossil-based electric generators, a higher fraction must be allocated to fossil generators to fairly offset the impacts of increased fuel prices. If, in the extreme, fossil generators were to bear all program costs passing nothing along to rate payers, they would need 40 percent of the total allocation pool to offset their costs. Therefore, 10-40 percent of the total allocation reflects the theoretical range of allowances needed to offset the financial impact of increased fuel prices in the electric sector.

Key Questions

- Should electricity generators be included in the allocation if they are not regulated?
- What portion of the total allocation should be granted to the electric power sector? Should it be based on the industry's share of greenhouse gas emissions or some other factor?
- Should generators in competitive and cost-of-service markets be treated differently under an allocation scheme?

- How should permits or allowances be distributed within the electric sector? Should it be based on historic emissions? Electricity output? Heat input?

g. Allocations for Energy-Intensive Industries?

Energy-intensive industries, such as steel, aluminum, chemicals, pulp and paper, and cement, would not be directly regulated in an upstream trading system. Like electric generators, these industries would, however, face higher prices for fossil fuels under a greenhouse gas trading system. While price increases would be modest, these industries consume significant amounts of fossil fuels and often face stiff competition from foreign competitors, most of whom would not be subject to mandatory greenhouse gas regulation. Including these industries in the allocation would not affect their incentive to improve efficiency and reduce fuel use, but it would offset increased energy costs and help to address competitiveness concerns associated with a domestic greenhouse gas trading program.

Without identifying exactly what businesses might be entitled to free permits or allowances, it is difficult to estimate a share of the overall allocation for this sector. If one provided allocations of free allowances only to the large, energy-intensive industries noted above—steel, aluminum, chemicals, and pulp and paper—close to 10 percent of the overall allocation would be required.

Key Questions

- Is there a sufficient policy rationale to have an allocation to selected energy- intensive industries? What industries should be included in the allocation?
- What portion of the overall allocation framework should be reserved for these industries?
- What are the appropriate metrics for determining allocations across different industries?

h. Allocations to other Industries/Entities?

Key Questions

- What other industries/entities (e.g. agriculture, small businesses, etc.) should be considered in the allocation pool?

- What should be the basis for their share of the total allocation as well as for the distribution among such industries/entities?

3. Should a U.S. System be Designed to Eventually Allow for Trading with other Greenhouse Gas Cap-and-Trade Systems being Put in Place around the World, such as the Canadian Large Final Emitter System or the European Union Emissions Trading System?

A greenhouse gas program in the U.S. could be designed to leave open the possibility of trading with greenhouse gas systems in other countries. There are both potential opportunities and challenges that arise with this type of linkage. On the positive side, numerous studies have shown that a trading system that includes emission reductions in key developing countries such as China and India will have significantly lower costs than a system that excludes these low cost reductions. Although links to trading programs in Europe and other developed countries are less beneficial from a cost standpoint, these links could nevertheless reduce costs and could facilitate efficient emission reductions within and between companies with operations in multiple countries

On the other hand, linkage to programs in Europe and other developed countries also raises several difficulties. Differences in design could complicate implementation and could lead to inconsistencies in allocation methods, monitoring and verification, or other design elements. In addition, disparities in the stringency of targets and in allowance prices could make linkage politically difficult.

Key Questions

- Do the potential benefits of leaving the door open to linkage outweigh the potential difficulties?
- If linkage is desirable, what would be the process for deciding whether and how to link to systems in other countries?
- What sort of institutions or coordination would be required between linked systems?

4. If a Key Element of the Proposed U.S. System is to "Encourage Comparable Action by other Nations that are Major Trading Partners and Key Contributors to Global Emissions," should the Design Concepts in the NCEP Plan (i.e., to Take some Action and then Make further Steps Contingent on a Review of what these other Nations Do) Be Part of a Mandatory Market-Based Program? If so, how?

Climate change is a global environmental problem that requires action by all major emitting countries. Participation by all key emitters is critical for two reasons. First, only with a global effort will it be possible to make sufficient progress to address the potential effects of climate change. Second, without greenhouse gas mitigation efforts by all major emitters, including our largest trading partners, the U.S. economy could be placed at a competitive disadvantage. Thus, an important component of a U.S. program could be to encourage major trading partners and large emitters of greenhouse gases to take actions that are comparable to those in the U.S. As noted above, some key developed countries, such as those in the European Union, are already implementing emissions trading programs. Other countries have developed efficiency standards and additional policies that reduce energy use and greenhouse gas emissions.

Key Questions

- What metrics are most valuable for comparison of developed and developing country mitigation efforts to U.S. efforts?
- What process should be used to evaluate the efforts of other nations and how frequently should such an evaluation take place?
- Are there additional incentives that can be adopted to encourage developing country emission reductions?

INDEX

A

accounting, 68
acid, 18, 22, 63
adaptation, 45, 64, 66
adjustment, 47
administration, 62
administrative, 36, 38, 62, 63
agriculture, 51, 66, 67, 70
aid, 49
air, 9
alternative, 18, 20, 41, 42, 43, 55
alternative energy, 43, 55
alternatives, 2, 14, 18, 24
aluminum, 35, 50, 70
amendments, 5
analysts, 8, 42
argument, 14
assessment, 2, 24
assets, 41, 49
assumptions, 7, 18, 20
atmosphere, vii, 13, 33, 56, 59, 60
automobiles, 37, 62
availability, 5, 6, 11, 19, 26

B

banking, 2, 3, 17, 18, 20, 21, 22, 23, 24
barriers, 69
behavioral change, 45
benefits, 2, 4, 5, 7, 8, 13, 16, 19, 20, 26, 27, 36, 43, 44, 51, 52, 53, 55, 57, 71
bifurcation, 16
borrowing, 2, 3, 18, 22, 23, 24
Btus, 47
bubble, 11
buildings, 37
burn, 49
Bush Administration, 2, 23

C

Canada, 31
caps, 2, 3, 4, 9, 18, 43
carbon, vii, viii, 2, 3, 4, 5, 6, 8, 7, 9, 10, 11, 12, 13, 14, 15, 16, 17, 19, 20, 23, 24, 26, 29, 30, 32, 34, 35, 36, 37, 38, 39, 40, 43, 44, 45, 46, 47, 49, 50, 52, 53, 54, 55, 56, 57, 59, 61, 63, 64, 67, 69
carbon dioxide, 2, 3, 10, 11, 14, 29, 34, 36, 38, 39, 46, 52, 53, 54, 56, 61, 63, 67
carbon emissions, 14, 34, 36, 37, 38, 39, 40, 43, 45, 47, 50, 55, 56, 63, 69
carbon monoxide, 15
cement, 35, 50, 70
CH4, 29, 30
chemicals, 5, 50, 70
China, 52, 71
circulation, 60
Clean Air Act, 4, 14, 16, 18, 31

Clean Development Mechanism, 11, 12
cleaning, 62
climate change, vii, 1, 2, 3, 6, 7, 8, 11, 13, 14, 20, 23, 24, 32, 60, 62, 64, 65, 66, 72
coal, 12, 16, 35, 37, 38, 40, 41, 47, 49, 62, 67, 68
coal mine, 62
combustion, 34, 36, 38, 47, 54
compensation, 35, 39, 40, 41, 48, 49
competition, 21, 41, 49, 70
competitive advantage, 64
competitive markets, 49
competitiveness, 6, 10, 23, 40, 50, 51, 60, 70
complement, 65
complex interactions, 20
complexity, 16, 20, 21, 24, 34, 47, 64, 68
compliance, 2, 3, 5, 9, 11, 14, 16, 17, 18, 20, 21, 24, 27, 30, 64, 67, 68
complications, 52
components, 66
concentration, 14, 57
confidence, 22
Congress, vii, 1, 24, 33, 59, 60
Congressional Budget Office, 29, 33, 34, 51, 53, 54, 55, 56, 57
consensus, vii, viii, 14, 59, 60
construction, 18
consumers, 2, 7, 19, 35, 40, 41, 45, 49, 50, 62, 64, 66
consumption, 34, 35, 37, 45, 47, 54, 55, 62, 68
control, 3, 4, 6, 10, 11, 13, 15, 16, 18, 20, 24, 64
correlation, 7
cost analyses, 8
cost saving, 18, 27
cost-effective, 4, 6, 11, 12, 14, 17, 19, 20, 21, 27, 36, 38, 51
costs, vii, viii, 1, 2, 3, 4, 5, 7, 8, 9, 10, 11, 12, 13, 14, 16, 17, 18, 19, 20, 21, 22, 23, 24, 26, 27, 28, 33, 34, 35, 36, 37, 38, 39, 40, 41, 42, 43, 44, 45, 46, 47, 48, 49, 50, 51, 52, 53, 54, 55, 56, 62, 64, 67, 68, 69, 70, 71
cost-sharing, 43, 65
covering, 37, 63

credit, 2, 6, 11, 15, 18, 19, 21, 22, 23, 24, 46, 67
credit market, 2, 11, 22, 23, 24
customers, 40, 49

D

Dallas, 29, 31, 32, 54, 56
damping, 16, 28
danger, 13
debt, 35, 41
decision makers, 2, 24
decisions, 14, 18, 35, 39, 40, 42, 43, 44, 48, 49, 50, 55, 56, 61
Denmark, 29
dependent variable, 13
designers, 24
developed countries, 23, 52, 71, 72
developing countries, 23, 52, 71
diesel, 38
disputes, 14
distortions, 12, 35, 55
distribution, vii, 1, 3, 5, 24, 26, 28, 40, 55, 62, 65, 71
division, 45, 66
droughts, 60
duration, 6

E

ecological, 4, 8
economic activity, 35, 41
economic growth, 9, 10, 19, 23, 26, 35, 55
economic incentives, 52
economics, 22, 26
election, vii, 1, 24
electric power, 5, 48, 69
electric utilities, 10, 40, 62, 64
electricity, 10, 15, 31, 34, 37, 38, 40, 41, 46, 47, 48, 49, 50, 54, 69
emission, 36, 37, 38, 42, 46, 50, 52, 53, 57, 63, 64, 65, 67, 71, 72
emitters, 4, 17, 18, 34, 38, 62, 72

energy, 9, 14, 15, 16, 18, 19, 37, 41, 43, 44, 45, 50, 51, 54, 55, 60, 62, 63, 64, 65, 66, 70, 72
energy efficiency, 37, 43, 55
Energy Information Administration, 54, 62
Energy Policy Act of 2005, 60
Environmental Protection Agency, 6, 31
estimating, 18, 27
estimators, 18
Europe, 71
European Union, 11, 30, 34, 51, 52, 71, 72
exports, 54
exposure, 66
extraction, 62

F

fairness, 62
February, 29, 31, 33
federal funds, 44
federal government, 4, 35, 43, 65
Finland, 29
firms, vii, 33, 34, 35, 37, 39, 40, 42, 43, 44, 46, 47, 48, 49, 50, 51, 52, 56, 67, 68
flexibility, 1, 5, 7, 10, 15, 16, 17, 18, 19, 22, 24, 27
flood, 65
flow, 13, 22, 43, 65
fluctuations, 10, 11, 26
forecasting, 7
forestry, 67
forward market, 22
fossil, 34, 36, 37, 38, 40, 46, 47, 48, 54, 55, 62, 66, 67, 68, 69, 70
fossil fuel, 34, 36, 37, 38, 40, 46, 47, 48, 54, 55, 62, 66, 67, 68, 69, 70
fossil fuels, 34, 36, 37, 47, 54, 68, 70
fuel, 4, 12, 36, 37, 38, 40, 45, 46, 47, 48, 52, 55, 62, 63, 66, 67, 68, 69, 70
funding, 12, 14, 17, 19, 22, 26, 42, 44, 45, 47, 65, 66
funds, 21, 43, 44, 45, 65, 66
furnaces, 62

G

gas, 7, 12, 30, 36, 37, 38, 41, 47, 54, 61, 62, 63, 64, 67, 68, 70, 71, 72
gases, vii, 1, 5, 13, 24, 29, 33, 61
gasoline, 38, 40, 63
general knowledge, 44
generation, 19, 31, 37, 38, 41, 49, 62, 69
generators, 15, 38, 46, 47, 48, 49, 63, 69, 70
global warming, 60
goals, 19, 67
goods and services, 34, 37, 40, 45
government, 4, 7, 12, 17, 20, 35, 39, 40, 41, 42, 43, 53, 55, 61, 64, 65, 67
government expenditure, 41
greenhouse, i, iii, v, vii, viii, 1, 2, 3, 6, 7, 8, 10, 11, 12, 13, 14, 17, 19, 22, 24, 26, 28, 29, 30, 31, 33, 34, 35, 36, 39, 43, 46, 48, 49, 51, 53, 54, 56, 59, 60, 61, 62, 63, 64, 65, 67, 69, 70, 71, 72
greenhouse gas (GHG), vii, viii, 1, 2, 3, 7, 8, 10, 11, 12, 13, 14, 17, 22, 24, 33, 34, 35, 36, 39, 43, 46, 48, 49, 51, 54, 56, 59, 60, 61, 62, 63, 64, 65, 67, 69, 70, 71, 72
greenhouse gases, vii, 1, 2, 3, 8, 10, 11, 12, 13, 14, 17, 22, 24, 33, 39, 43, 54, 56, 59, 60, 61, 62, 72
gross domestic product, 41
grouping, 15
groups, vii, 33
growth, vii, 1, 3, 9, 10, 19, 20, 23, 24, 26, 35, 55, 60
guidance, 15

H

hardness, 21
harm, vii, 1, 3, 4, 24, 60
heat, 12
higher-income, 45
host, 15, 19
household, 45
household income, 45
households, 34, 37, 40, 45, 49, 50, 62, 64, 66

human, 4, 60, 65
human activity, 60
hybrid, 2, 13, 15, 17, 20, 22, 24
hydro, 41
hydrogen, 43

I

implementation, 1, 3, 7, 15, 17, 22, 24, 25, 37, 38, 47, 67, 71
incentive, vii, 10, 17, 20, 33, 34, 36, 37, 38, 43, 44, 47, 49, 50, 55, 64, 70
incentives, vii, viii, 1, 2, 3, 24, 34, 36, 37, 38, 41, 43, 44, 46, 47, 49, 55, 59, 60, 64, 66, 72
inclusion, 7, 36, 42, 53
income, 41, 45, 55, 64, 66
India, 52, 71
indication, 22
indices, 10
industrial, 37, 38, 61, 62
industrial sectors, 38, 61
industry, 5, 6, 10, 47
innovation, 12, 17, 37, 38, 42, 44, 65
insight, 39
institutions, 51, 71
instruments, 21
integrity, 22
intensity, 9, 10, 20
interaction effect, 55
interactions, 20
interference, 60
inventories, 5
investment, 14, 19, 43, 44, 69

J

jobs, 48
justification, 43

K

Kyoto Protocol, 2, 11, 23, 29

L

labor, 35, 41, 42, 55
land, 40
landfills, 67
language, 2
layoffs, 2
leakage, 38
legislation, 7, 12
legislative proposals, 38
lenses, 8
licenses, 4
life-cycle, 8, 61
life-cycle cost, 8
lifetime, 13, 63
LIHEAP, 66
linkage, 51, 71
links, 71
liquidity, 5, 18
local government, 55
Los Angeles, 15, 29
lower-income, 45
low-income, 45, 64, 66

M

mandates, 2, 7, 10, 14, 16
manipulation, 16
manufacturing, 63, 68
marginal cost curve, 4
marginal costs, 12, 48
market, vii, viii, 1, 2, 3, 4, 6, 7, 8, 9, 11, 13, 15, 16, 17, 18, 19, 20, 21, 22, 23, 24, 26, 27, 28, 34, 41, 43, 49, 52, 55, 59, 60, 64
market disruption, 18
market economics, 8
market failure, 22
market stability, 7
marketing, 21
marketplace, 2, 3, 13, 19, 28
markets, 7, 10, 11, 23, 48, 49, 50, 69
maximum price, 36, 52
measures, 9, 13, 64, 65
melting, 65

methane, 13, 29, 30, 67
metric, 6, 46, 54, 56
mining, 62, 67
missions, 3, 30, 34, 38, 51, 52, 63, 67, 68
modeling, 8
models, 5, 7
money, 4, 14
motion, 29, 32

N

nation, 67
natural, 12, 37, 38, 41, 47, 59, 62, 65, 67, 68
natural gas, 12, 37, 38, 41, 47, 62, 67, 68
Netherlands, 29
New York, 31
nitrogen, 15
nitrogen oxides, 15
nitrous oxide, 29, 30
non-profit, 43, 65
Norway, 29
nuclear, 41, 44

O

observations, 39, 48
oil, 38, 39, 40, 55, 62
oil refineries, 62
organizations, 39, 67
oxide, 15, 29, 30
ozone, 5, 13

P

paper, 3, 31, 32, 33, 34, 36, 50, 51, 70
penalty, 24
performance, 19, 63
periodic, 5
permafrost, 65
permit, 2, 3, 4, 11, 13, 14, 15, 16, 17, 20, 62, 65, 68
petroleum, 12, 62, 63, 67, 68
pipelines, 62, 67
planning, 16, 18

plants, 62
play, 11, 43, 45
policymakers, vii, 19, 33, 34, 35, 40, 41, 42, 46, 49, 51, 53, 57
polluters, 4
pollution, 4, 13, 15, 18
power, 5, 6, 15, 16, 41, 48, 49, 69
power plants, 15
precipitation, 65
pre-existing, 19
preference, 2, 4
President Bush, 29
President Clinton, 15
pressure, 16
price changes, 40
price floor, 19, 26
price index, 13, 41
price mechanism, 15, 22
price stability, 27
prices, 7, 9, 11, 13, 15, 18, 20, 21, 22, 27, 28, 35, 37, 40, 41, 43, 45, 47, 49, 50, 52, 54, 55, 63, 66, 68, 69, 70, 71
private, 43, 44, 67
probability, 22, 23
producers, 7, 37, 40, 41, 42, 47, 48, 62, 67, 68
production, 5, 10, 34, 35, 36, 37, 42, 48, 49, 50, 51, 54, 55, 56, 62, 67, 68, 69
production costs, 50, 69
profit, 43, 44, 65
profitability, 11
profits, 35, 44, 47, 50, 51, 55, 64
program, vii, viii, 1, 2, 3, 4, 5, 6, 7, 9, 10, 11, 13, 14, 15, 16, 17, 18, 19, 20, 21, 22, 23, 24, 26, 27, 30, 31, 33, 34, 35, 36, 37, 38, 39, 40, 41, 42, 43, 44, 45, 46, 47, 48, 49, 50, 51, 52, 53, 54, 55, 56, 59, 60, 61, 62, 63, 64, 66, 67, 68, 69, 70, 71, 72
program administration, 62
promote, 16, 22, 23, 43, 65
property, 66
prosperity, 60
protection, 22, 26
public, vii, 6, 19, 33, 65
public interest, vii, 33
public sector, 65

public support, 6

R

R&D investments, 44
radical, 12
rain, 18, 22, 63
range, 3, 7, 8, 11, 17, 37, 53, 59, 62, 69
rate of return, 14
real income, 55
reduction, vii, 1, 2, 3, 4, 5, 6, 7, 9, 10, 11, 13, 14, 15, 16, 17, 18, 21, 22, 24, 26, 27, 34, 38, 46, 51, 52, 53, 56, 62, 63, 65, 67
refineries, 68
refiners, 62, 63, 67, 68
regional, 15, 26, 45, 66
regulation, 33, 34, 39, 48, 62, 67, 70
regulations, 16, 35
regulators, 16, 49, 54
renewable energy, 37
research, 6, 13, 14, 16, 20, 22, 23, 37, 43, 44, 45, 57, 65, 66
research and development, 6, 14, 16, 22, 23, 37, 44
researchers, 39
residential, 38, 62
resolution, vii, 1, 2, 24, 60
resources, 17, 42, 65
responsiveness, 16, 27
returns, 41
revenue, 12, 13, 22, 23, 35, 41, 42, 44, 49, 64, 65, 69
risk, 4, 10, 11, 16, 17, 60
risks, 14, 26, 66

S

safeguards, 26
safety, 2, 11, 13, 18, 19, 20, 21, 22, 23, 24, 28, 36, 42, 52, 53, 54, 56
sales, 68
sample, 3
savings, 8
sea level, 60, 65

Senate, vii, viii, 1, 2, 24, 59, 60
services, 34, 37, 40, 41, 45
severity, 60
shareholders, 35, 48, 51, 55
sharing, 43, 65
shocks, 18, 66
short period, 43
shortage, 10
short-term, 7, 10, 13, 16, 17, 18, 26, 27
signals, 2, 16, 24
smog, 13, 15
smokestacks, 38
smoothing, 18
Social Security, 41
solar, 16, 43, 44
specificity, 25
spot-market, 15
stability, 5, 7, 19, 23, 27
stabilization, 14
stabilize, 23, 30
stages, 13
standards, 72
statutory, 4
steel, 50, 70
stock, 13, 14
storage, 38
strategies, 3, 66
structural changes, 17
subsidies, 43
substitution, 7, 54
sulfur, 4, 16, 29, 30
sulfur dioxide, 4, 16
supplemental, 22
suppliers, 34, 40, 47, 62
supply, 11, 13, 22, 23, 35, 40, 48, 55
Sweden, 29
systems, vii, viii, 15, 19, 21, 34, 51, 52, 59, 62, 63, 65, 71

T

targets, 3, 10, 15, 16, 21, 27, 52, 62, 69, 71
tax credit, 43, 44, 65
tax credits, 43, 44, 65
tax receipt, 41

taxation, 42
taxes, 2, 4, 5, 7, 12, 19, 24, 35, 41, 55
technological change, 7
technological progress, 18
technology, vii, viii, 1, 2, 3, 5, 6, 7, 8, 10, 12, 13, 14, 16, 17, 19, 20, 21, 22, 23, 24, 26, 27, 28, 42, 43, 44, 59, 64, 65
Third World, 11
threatened, 12
threshold, 15, 57
threshold level, 57
time, 9, 11, 14, 15, 16, 17, 18, 20, 21, 22, 27, 35, 43, 45, 46
time frame, 9, 14, 16, 21
timetable, 5, 16, 17, 27
timing, 5, 6
trade, vii, 2, 3, 4, 13, 17, 20, 22, 30, 33, 34, 35, 36, 37, 38, 39, 40, 41, 42, 43, 44, 45, 46, 47, 48, 49, 50, 51, 52, 53, 55, 56, 64
tradeable permits, 4
trading, vii, 4, 5, 7, 9, 11, 12, 13, 15, 17, 18, 20, 22, 24, 26, 31, 33, 34, 46, 49, 51, 52, 60, 62, 63, 64, 67, 68, 69, 70, 71, 72
trading partners, 34, 60, 72
trajectory, 9, 10
transfer, 41
transfer payments, 41
transition, 6, 13
transparent, 12
transportation, 5, 37, 38, 62, 63
trees, 47
trend, 16
triggers, 11, 15, 16, 27
trucks, 63
turnover, 7, 14, 18

U

U.S. economy, 62, 64, 72
uncertainty, vii, 1, 2, 3, 4, 5, 7, 10, 11, 13, 14, 16, 17, 18, 20, 24, 26, 27
unit cost, 2, 3
United Nations, 2, 23
United States, vii, 1, 2, 5, 11, 24, 33, 37, 52, 54, 55, 60, 61, 62, 63

V

variability, 59
variable, 2, 3, 9, 12, 13, 16, 26, 27
variables, 5, 6, 7, 10, 15, 26
vehicles, 63
vein, 14
voice, 29, 32
volatility, 7, 13, 16, 17, 18, 20, 21, 27, 28

W

wages, 35, 47, 50
wells, 39, 62
wind, 43, 44
workers, 2, 35, 40, 48, 51
World Resources Institute, 8, 29, 30, 31

Y

yield, 4

Z

zoning, 65